D1810139

Where Is Everyone?

This is a game to play on your own.

How to play

- You will need a die. The number thrown on the die is the number of people living in any square of the grid below.
- Throw the die and write down the number in the space headed 'Number thrown' in the databank.
- Throw the die 12 times. Each time record the number thrown.
- Use the databank information to shade each square of the grid using the following key:

    Squares with **5 or 6** people are **crowded**. Shade these **red**.
    Squares with **3 or 4** people are **less crowded**. Shade these **orange**.
    Squares with **1 or 2** people are **uncrowded**. Shade these **yellow**.

Databank 1		
Throw	**Square on grid**	**Number thrown**
1	0072	
2	0073	
3	0074	
4	0172	
5	0173	
6	0174	
7	0272	
8	0273	
9	0274	
10	0372	
11	0373	
12	0374	

Grid for databank 1

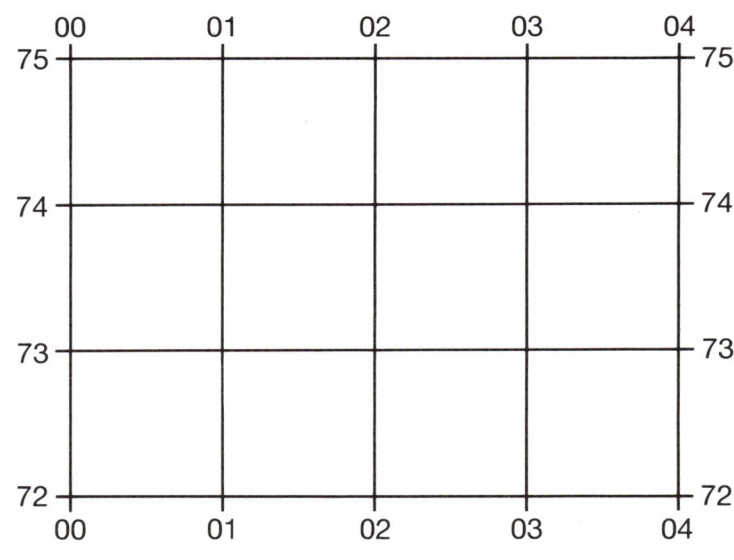

1 Write down the grid reference of two squares which are **crowded**.

Square_____ and square_____

2 Write down the grid reference of two squares which are **less crowded**.

Square_____ and square_____

3 Write down the grid reference of two squares which are **uncrowded**.

Square_____ and square_____

4 Are there more **crowded** squares than **uncrowded** squares? *Circle* YES/NO/SAME

Your coloured-in grid shows that people are **not evenly spaced in any area**.

Where Should We Live?

This is a game for two players.

The winner is the player who gains most squares of industrial land.

Look at the databank on this page. It has four columns.
Column **1** shows the **references to the squares** on the grid.
Column **2** shows the **type of land** in the squares on the grid.
Column **3** is for you to record the **number thrown** on the die.
Column **4** has 20 boxes for recording **each throw** you make.

Databank 2

Columns			
1	**2**	**3**	**4**
Square on the grid	Type of land	Number thrown	Throws made
0072	fertile		
0073	fertile		
0074	fertile		
0172			
0173			
0174			
0272	desert		
0273	desert		
0274			
0372	desert		
0373	desert		
0374	industry		

Grid for databank 2

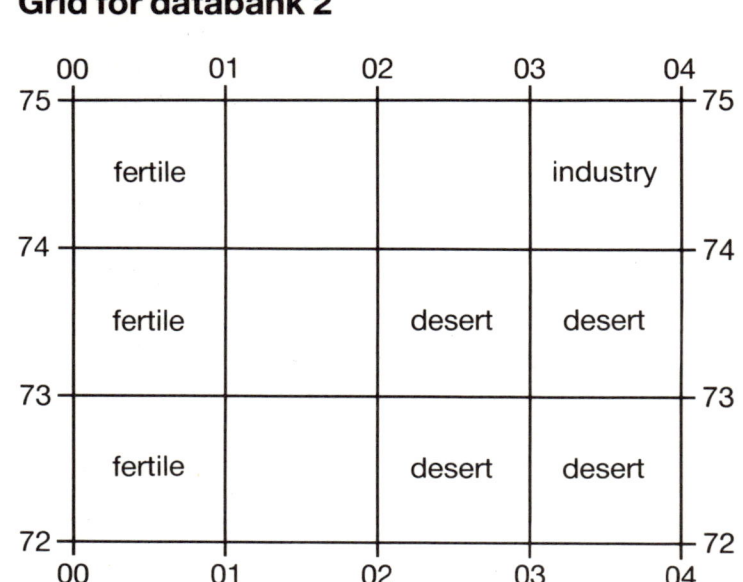

How to play

● You will need a die. Take turns to throw it.
● Each player has **up to 20 throws** of the die.
● After each throw, tick a box in column 4 and decide where, in column 3, to write the number.
● For **industrial land**, you must throw **5 or 6**. Write 5 or 6 against 'Industry' in column 3.
● For **fertile land**, you must throw **3 or 4**. Write 3 or 4 against 'Fertile' in column 3.
● For **desert land**, you must throw **1 or 2**. Write 1 or 2 against 'Desert' in column 3.
● You will notice that four squares are blank. **You can decide the type of land** in these squares from the numbers you throw.

In the grid above, three types of land are shown:

☐ Land used for industry. Usually, more people live there than live on fertile farmland or in the desert. This land has a **high population density**.

☐ Fertile farmland. Usually, more people live there than live in the desert. This land has a **medium population density**.

☐ Desert. Usually, few people live there. This land has a **low population density**.

Where people live

More than four billion people (that is 4 000 000 000) live in the world. They are not evenly spread out. Some areas of the world are more crowded than others.

You will see this when you have completed the map on page 4.

How to fill in the map

● The map shows some areas of the world where people live.
● Each area has been labelled A, B, or C.
  The areas marked **A** are **least crowded.**
  The areas marked **B** are **not very crowded.**
  The areas marked **C** are **most crowded.**
● Colour the map as follows:
  Shade the **A** areas **yellow**.
  Shade the **B** areas **orange**.
  Shade the **C** areas **red**.
  Leave the areas without letters in them **blank**.
● Complete the key on the map.

Now do this

1 Fill in the boxes below with the names of the countries numbered 1 to 14 on the map. (You may need to look at an atlas.) The names of the countries are given below.

Names of the countries 1 to 14
Argentina, Australia, Brazil, Canada, China, India, Japan, Mexico, New Zealand, Nigeria, South Africa, United Kingdom, USA, USSR

2 How many of the areas are **most crowded**? Name these on the line below.

_____

3 Name **three** countries **west** of the Greenwich Meridian (0° longitude) which are **least crowded**.

  (1)_____ (2)_____ (3)_____

4 Name **two** countries **east** of the Greenwich Meridian which are **least crowded**.

  (1)_____ (2)_____

5 Name **two** countries **west** of the Greenwich Meridian (0° longitude) which are **not very crowded**.
  (1)_____ (2)_____

6 Name **four** countries **east** of the Greenwich Meridian which are **not very crowded**.
  (1)_____ (2)_____
  (3)_____ (4)_____

7 Are the most crowded areas **north** of the equator?   *Circle*   YES/NO

8 Are the most crowded areas **south** of the equator?   *Circle*   YES/NO

9 Are the most crowded areas mainly **west** of the Greenwich Meridian?   *Circle*   YES/NO

10 Are the most crowded areas mainly **east** of the Greenwich Meridian?   *Circle*   YES/NO

11 Write a sentence here to describe where the **most crowded** areas of the world are to be found.

_____

_____

_____

The map on page 4 shows that more people live in some parts of the world than in other parts. It is also the same in any country. More people live in some parts of a country than in others.

Think about the area in which you live. Near your home there may be many houses crowded together, with a lot of people living in them. There may be other areas quite near to your home which have fewer houses and fewer people living in them.

● Give the name of a **crowded** area near your home.

_____

● Give the name of an **uncrowded** area near your home.

_____

1		2		3		4	
5		6		7		8	
9		10		11		12	
13		14					

World map: where people live

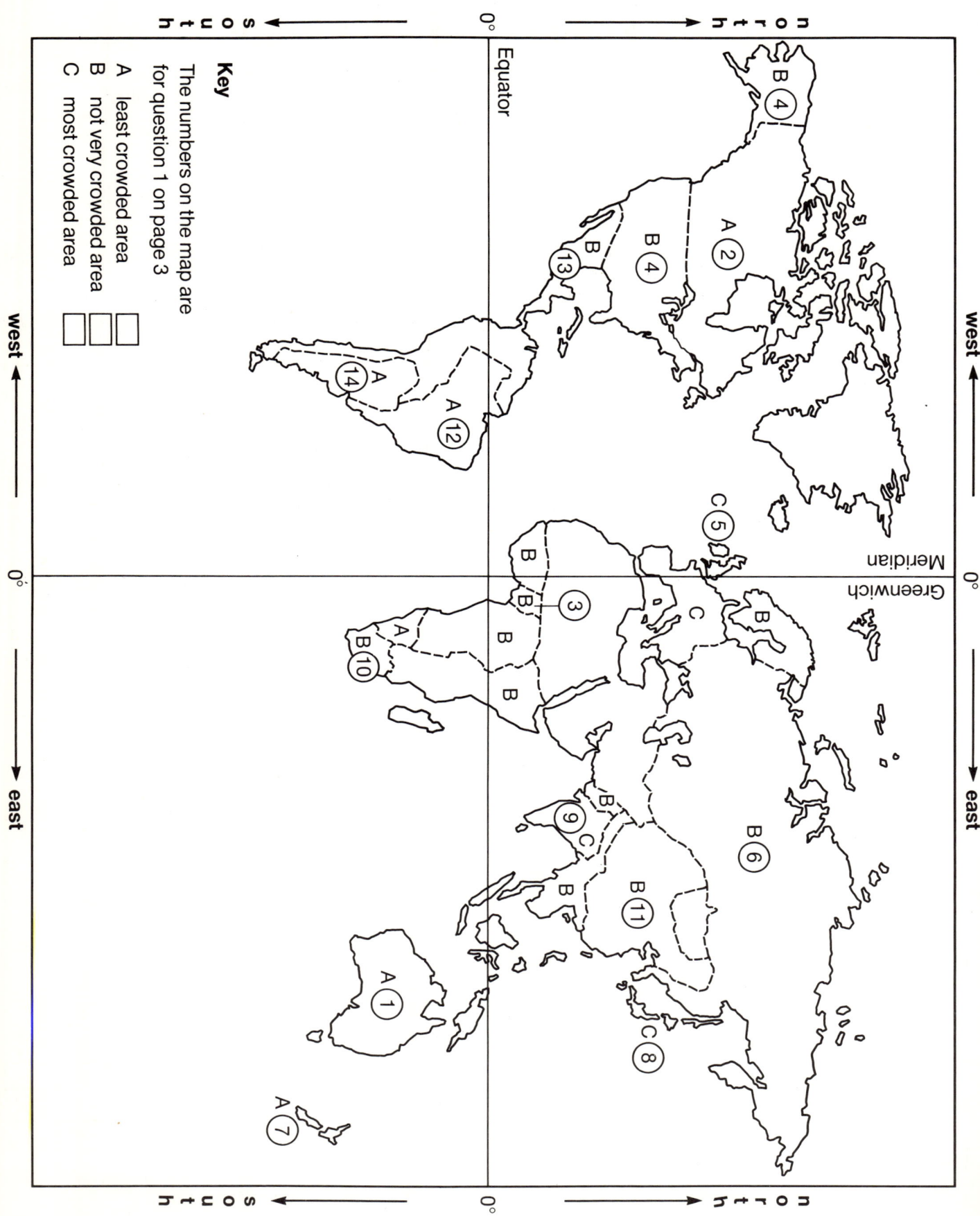

Key

The numbers on the map are for question 1 on page 3

A   least crowded area
B   not very crowded area
C   most crowded area

A ☐ B ☐ C ☐

Where people live in Great Britain

This map of Great Britain is divided up to show the counties.
- Use **yellow** to shade the counties marked **A**.
- Use **orange** to shade the counties marked **B**.
- Use **red** to shade the counties marked **C**.

1 How many **most crowded** counties are there? _____

2 Look at Scotland, England and Wales.
Which of them has the greatest number
of **most crowded counties**? _____

3 Are the **least crowded areas** in the highlands
or lowlands of Scotland and Wales? (You may
use maps in an atlas to help you.) _____

4 Name a **most crowded county**
which is also **a capital city**. _____

The map of the world (page 4) and this map
of Great Britain show **population density**.
A population density map shows how many
people live in a certain area. A high population
density means an area is crowded.

Some counties in Great Britain are not very
crowded overall, but may have crowded areas
within them. In the same way, some countries
on the world map are not very crowded overall
but have areas within them which are crowded.

A	least crowded area
B	not very crowded area
C	most crowded area

Where people live in North America

Here is an outline map of North America showing the boundaries between its countries.
● On the key beside the map, name the countries marked X, Y and Z.
● On the map, draw the **boundaries** in **red**.

When you completed the map of Great Britain on page 5, you could see that some parts were more crowded than others. The same is true of North America.

One reason why the population is not evenly spread out is that some areas are high mountains, some areas are hot deserts, some areas are forests, and some areas are too cold to live in. These areas are sometimes called **hostile** areas. Not many people live in them.

● To complete this map, you will first need to do some drawing and shading on page 7.
● When you have completed the map and the key, turn to page 8.

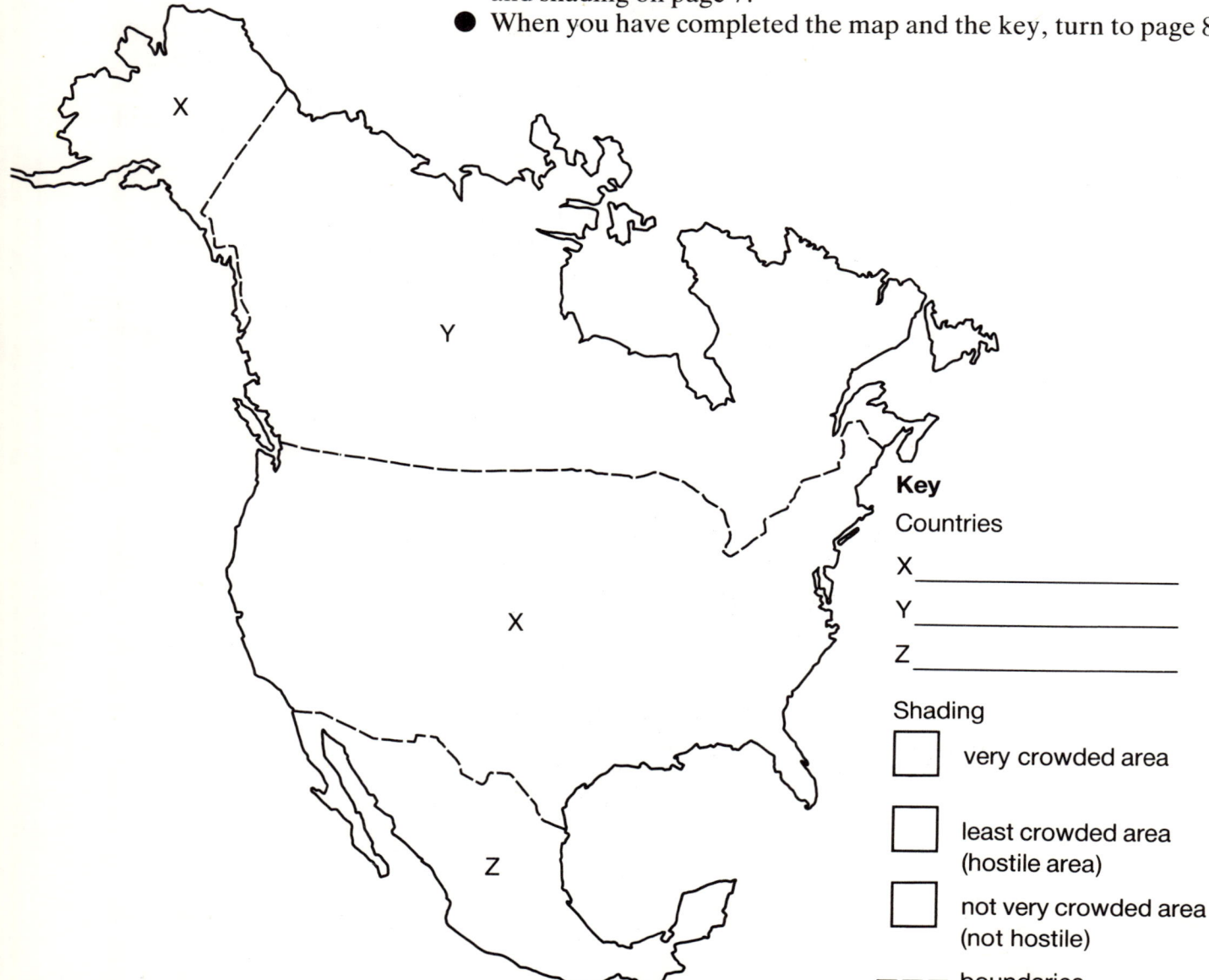

Key

Countries

X _____

Y _____

Z _____

Shading

☐ very crowded area

☐ least crowded area (hostile area)

☐ not very crowded area (not hostile)

‐ ‐ ‐ boundaries

Where people live in North America
(continued)

Here are some shapes. When they are **cut out**, they will fit together on the map of North America on page 6. The shapes show areas in the continent which are either **hostile** or **very crowded**. The rest of the continent is neither hostile nor very crowded. You will notice that the state **boundaries** are drawn on the shapes. These will help you to fit them in the correct places.

How to complete the map on page 6

● Draw the **boundaries** in **red**.
● Shade the **very crowded** area **red**.
● Shade the **hostile** areas **yellow**.
● Cut out the shapes carefully. Put them on the map on page 6 in their correct places. (Look at the boundaries. They will help you.)
● Draw around the edges of the shapes. Make sure they **stay in place** as you draw.
● Shade the areas on the map in the **same colours** that you used to colour the shapes.
● Complete the map by shading the remaining areas **orange**. The areas shaded orange are neither very crowded nor hostile.

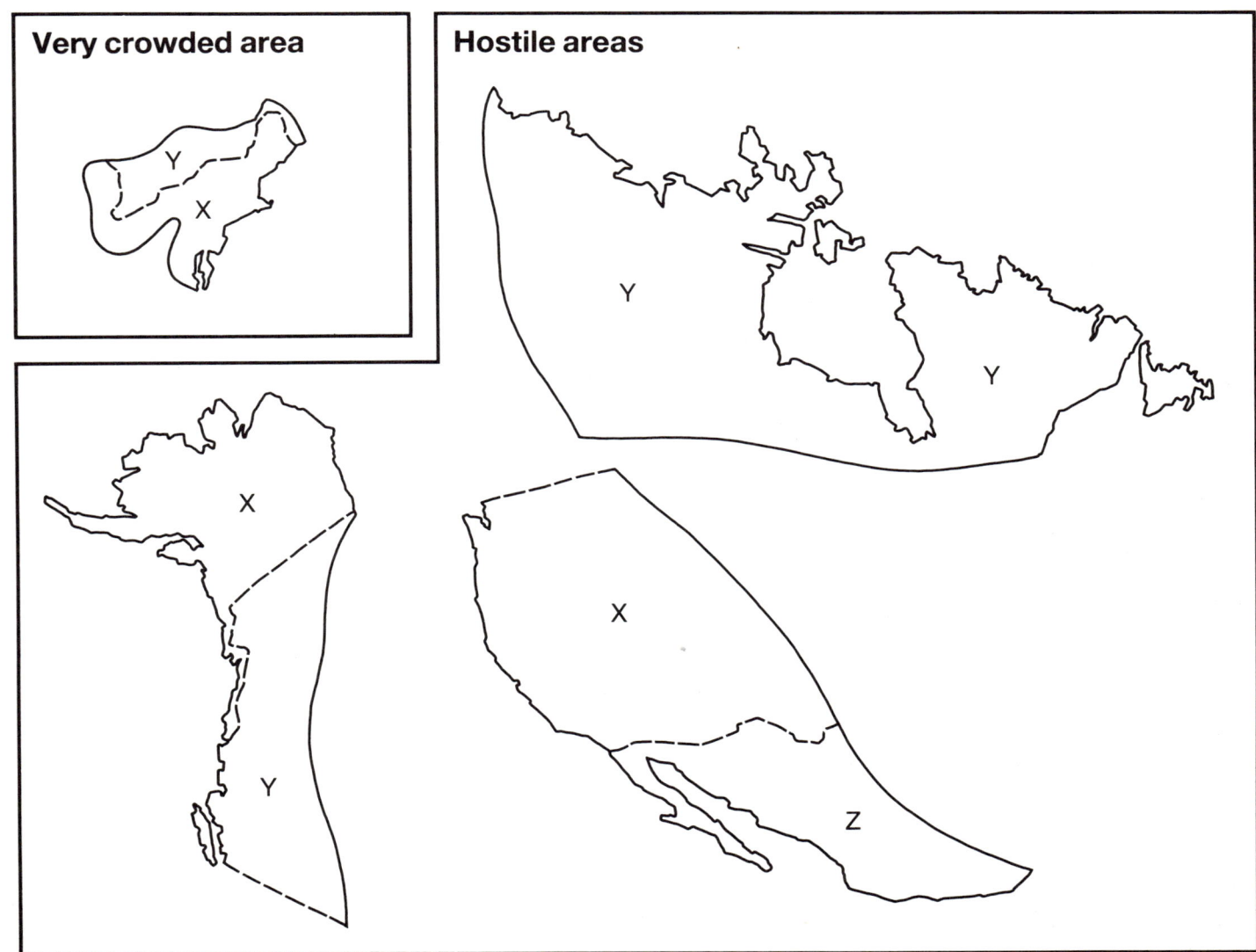

Very crowded area

Hostile areas

Where people live in North America (continued)

1 What are **hostile** areas like?_____

2 Name a **range of mountains** which form part of the hostile area of North America? (You may use an atlas.)_____

3 Name a city which is found in the **very crowded** area of North America? (You may use an atlas.)
_____

Heads and Tails

The sentences below have been split and mixed up. Try to work out the complete sentences by **matching** the **tails** with the **heads**.
When you are sure that you have matched them up, write them out on the sheet of paper your teacher will give you.

Heads

1 Most crowded areas . . .
2 Areas which are difficult to live in . . .
3 Those places where crops grow easily . . .
4 In Great Britain . . .
5 The Greenwich Meridian, 0° longitude, is a line on a map . . .

Tails

A . . . England is the most crowded of the three countries.
B . . . which is drawn from the North Pole to the South Pole.
C . . . are sometimes desert or mountain areas.
D . . . are called fertile areas.
E . . . are those which are good to live in.

Crossword

The number of dashes (− − −) in a clue show you how many letters there are in the word you are looking for.

Clues across

3 An − − − − − − − area is an area where not many people live.
5 More land is found − − − − − of the equator.
7 Cowboy films are usually about the Wild − − − −.
8 − − − − − − is a country in North America.
9 Great Britain is divided into − − − − − − − −.
11 More ocean is found − − − − − of the equator.
14 The smallest country in North America is − − − − − −.

Clues down

1 North America is a − − − − − − − − −.
2 The direction opposite to 7 across is − − − −.
4 The number of 'not very crowded counties' in Wales is − − −.
6 The initial letters of a North American country are − − −.
7 The smallest country in Great Britain is − − − − −.
8 On the maps, the areas shaded red are − − − − − − −.
10 The same as 11 across − − − − −.
12 North America has − − − − − countries.
13 New − − − − City.

If You Are . . .

This is a game for two players.

How to play

- You will each need a pencil and an eraser.
- Decide who will go first by tossing a coin.
- Take it in turn to read **each instruction**.
- You must do what each instruction says.
- You have to **shade** the boxes in the column.
- The winner is the player who has the highest column when **all 18 instructions** have been carried out.

Column

15
14
13
12
11
10
9
8
7
6
5
4
3
2
1

Instructions

1 If you are wearing trousers or a skirt, shade 2 squares.

2 If you are wearing a watch, shade 1 square.

3 If you are wearing an electronic watch, rub out 1 square.

4 If you are wearing a tie, shade 2 squares.

5 If you are not wearing a tie, rub out 1 square.

6 If you are wearing a blue, black or grey pullover, shade 1 square.

7 If you have dark hair, shade 1 square.

8 If you have fair hair, rub out 1 square.

9 If you have red or ginger hair, shade 2 squares.

10 If you have blue eyes, shade 1 square.

11 If you are older than your partner, shade 1 square.

12 If you are taller than your partner, shade 1 square.

13 If you are wearing brown or black shoes, not trainers, shade 1 square.

14 If you walked to school, shade 1 square.

15 If you came to school by car, rub out 2 squares.

16 If your name begins with a letter between A and J, shade 1 square.

17 If you wear glasses, shade 1 square.

18 If any of your teeth have been filled, rub out 3 squares.

Population of the United Kingdom

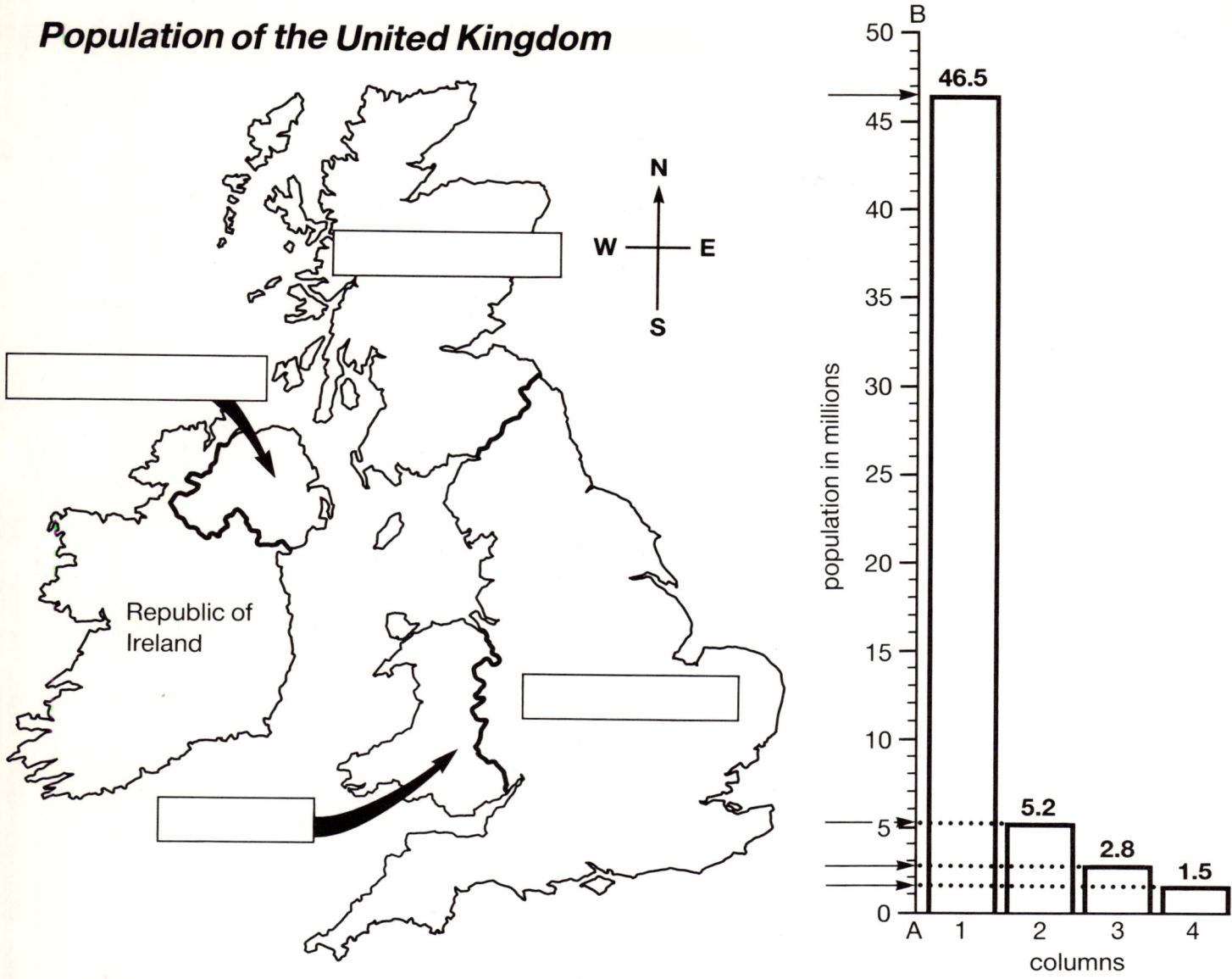

Line AB in the graph above right is called the **vertical axis** of the graph. In this graph, the vertical axis shows the number of people who live in the countries of the United Kingdom. These countries are England, Northern Ireland, Scotland and Wales.

● Write in the boxes on the map above the names of the four countries of the United Kingdom. (You may use an atlas to help you.)

There are four columns on the graph above right.
   Column **1** is **England**. Shade it **red**.
   Column **2** is **Scotland**. Shade it **blue**.
   Column **3** is **Wales**. Shade it **yellow**.
   Column **4** is **Northern Ireland**. Shade it **green**.

The number at the top of each column is the population of that country in millions.

● Colour the countries the same colours as those you used on the graph above.

1 Which country has the most people (the highest population)? _____

2 Which country has the fewest people (the smallest population)? _____

3 List the countries according to their population. Put the country with the highest population first, the country with the next highest population second, and so on, until you have completed the list of four.

1st _____   2nd _____

3rd _____   4th _____

Where people live

The table below shows approximately how many people live in each continent of the world.

Continent	Colour code	Population
Africa	brown	460 million
Asia (not including USSR)	yellow	2500 million
Australia and New Zealand	green	18 million
Europe (not including USSR)	purple	475 million
North America	red	335 million
South and Central America	orange	240 million
USSR (This is not a continent. It is a country in two continents.)	pencil	270 million

The figures in the table have been used to make a **bar graph** for each continent.

What to do

● Shade the bars in the colours shown in the table.

1 Which continent has the **greatest population**?

_____

2 Which two continents have populations of **similar size**?

_____

3 Which continent has the **smallest population**?

_____

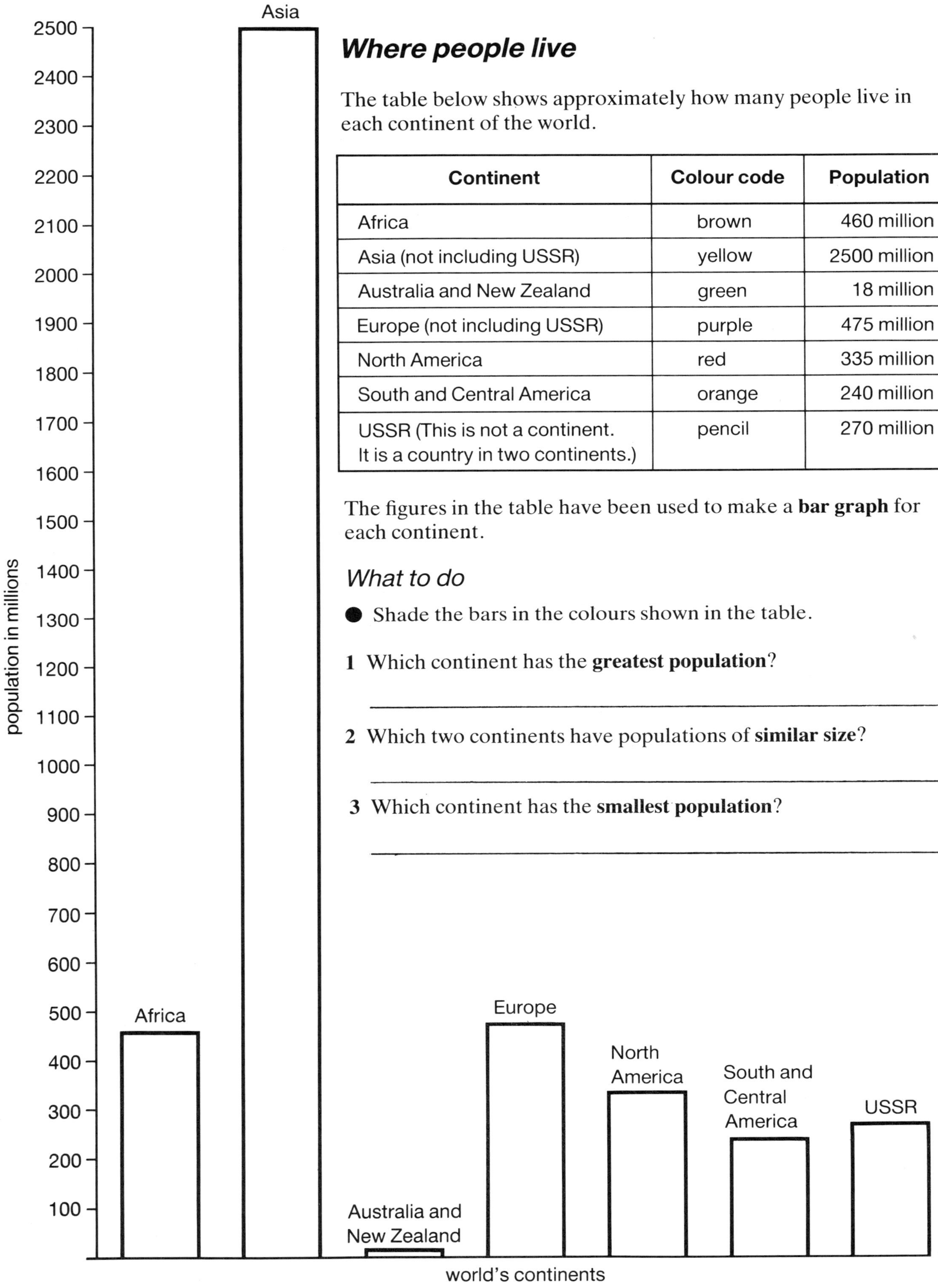

population in millions

2500 — Asia
2400 —
2300 —
2200 —
2100 —
2000 —
1900 —
1800 —
1700 —
1600 —
1500 —
1400 —
1300 —
1200 —
1100 —
1000 —
900 —
800 —
700 —
600 —
500 — Africa   Europe
400 —   North America
300 —   South and Central America   USSR
200 —
100 —
          Australia and New Zealand

world's continents

Populations of some countries

What to do

● Use an atlas to name each of the countries marked with dashes on the map on page 13. Each dash stands for one letter of the country's name. The names of the countries are given below the map on page 13. Write the names of the countries on the dashes. Brazil has been done for you.

● Fill in **databank 1** below.

▶ The databank shows the population for each country.
▶ The databank has a column called 'Height of bar'. Complete this column by filling in the correct height for each country's population.
▶ In the bar graphs you are going to draw on the map, 1 mm stands for 10 million people. For example, the population of Brazil is 130 million. This means that the bar will be:

$$130 \div 10 = 13 \text{ mm high}$$

Brazil has already been done for you.

Databank 1		
Country	**Population (million)**	**Height of bar (mm)**
Brazil	130	13
Britain	56	
China	1000	
India	750	
Japan	120	
Nigeria	90	
USA	235	
USSR	270	

● Complete the bars on the world map on page 13.
● Fill in **databank 2** below by writing the name of each country beside a population total which is correct for that country.

You have now listed the countries in rank order according to their populations.

Databank 2	
Country	**Population (million)**
	1000
	750
	270
	235
	130
	120
	90
	56

World map: populations of some countries

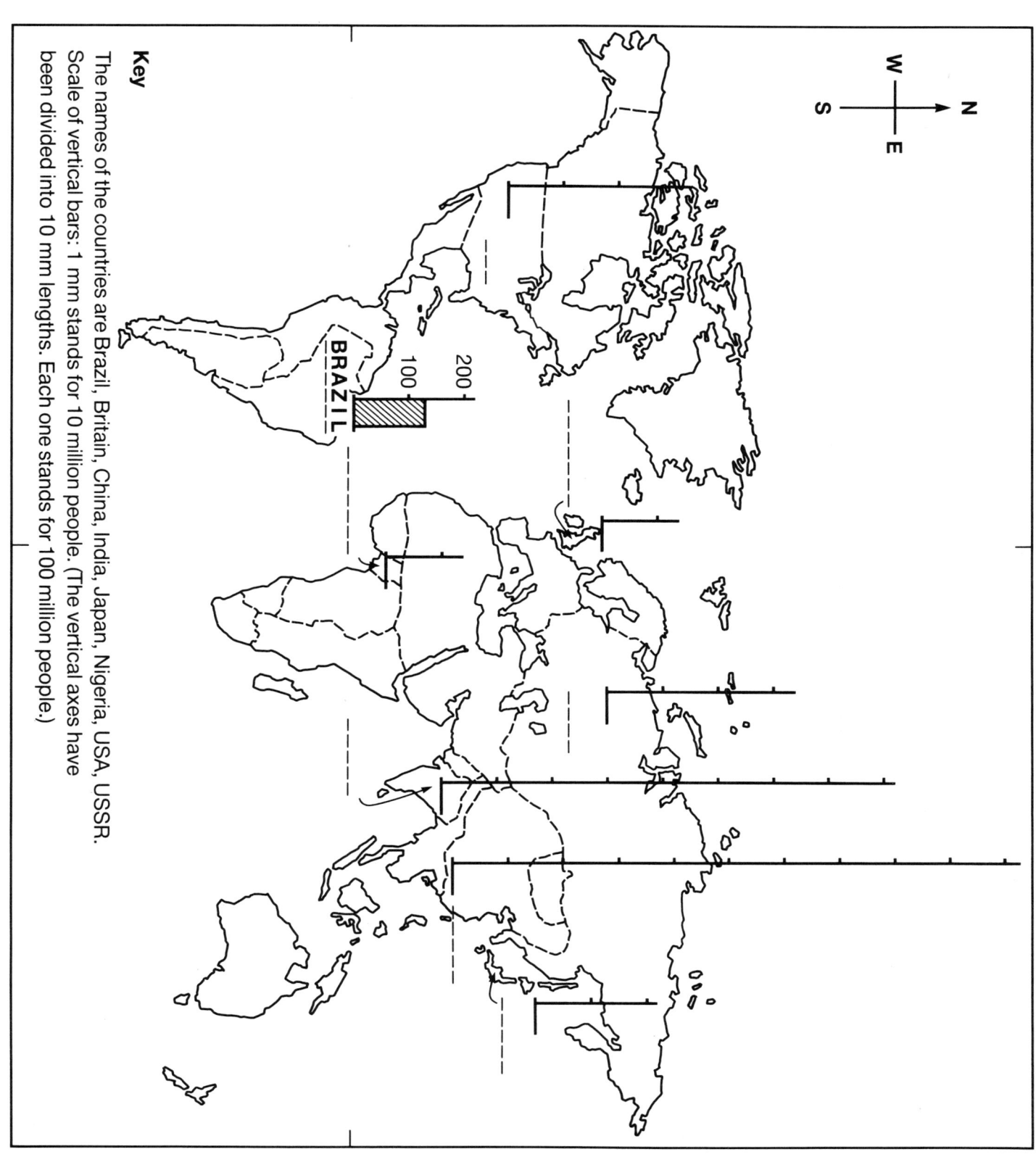

Key

The names of the countries are Brazil, Britain, China, India, Japan, Nigeria, USA, USSR.
Scale of vertical bars: 1 mm stands for 10 million people. (The vertical axes have been divided into 10 mm lengths. Each one stands for 100 million people.)

Where crops can be grown

Continent	Cropland area (million square km)
Africa	9.75
Asia	10.5
Australia	9.75
Europe	2.25
North America	3.5
South America	3.25

Each continent has land which is good for growing crops and land which is not.

The table on this page shows how much crop-growing land there is in each continent. The figures are approximate ones.

A bar graph can be drawn to show the same thing.

What to do

● Draw in the bars for the continents using the information in the table. Africa has been drawn for you.

● Shade the bars the same colours as you used for the continents on page 11.

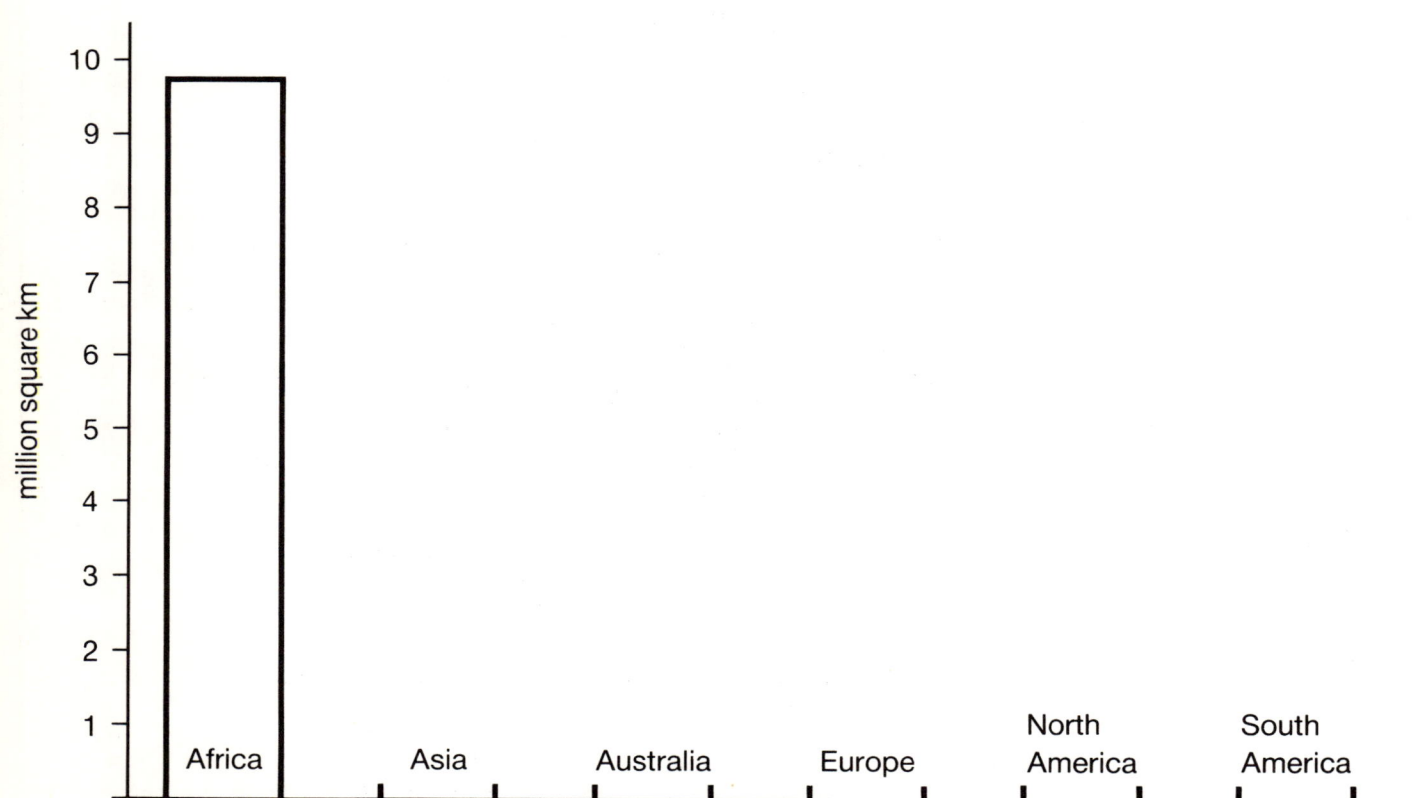

1 Which continent has the **greatest area** of cropland? _____

2 Which continent has the **smallest area** of cropland? _____

3 Name **two** continents which have the **same area** of cropland. _____

and _____

People and the land

Look carefully at the bar graph on page 11 and the bar graph on page 14.

1 Does the continent with the **largest area** of cropland have the **largest population**? *Circle*   YES/NO
2 Does the continent with the **smallest area** of cropland have the **smallest population**? *Circle*   YES/NO
3 Which continent has the **smallest area** of cropland?

_____

4 **How many people** live in the continent which has the **smallest area** of cropland? _____

5 Which **two** continents have **similar areas** of cropland?

(1)_____ (2)_____

6 Which continent has a **very small population** and cropland of more than **9 million square kilometres**?

_____

How many classrooms?

What to do

● On the back of this page make a large copy of this outline for a graph. It has **six** bars, one for each group of classrooms.

● Count up the number of classrooms in your school and put them into these groups:
   science, craft, technology and design, art, geography and history, maths, English

● On the grid which you have drawn, draw **six** bars to show the number of classrooms in each group. Name the bars to show which classrooms they stand for. A plan of your school will be helpful in doing this.

Constructagraph

This is a game for two, three or four players.

How to play

● Each player needs a column grid, a databank and a spinner. The grid is on page 17.
● Before you start, shade the boxes in the databank below in the **same colours** as those on the spinner.
● Decide the order of players. Then spin the spinner in turn.
● See which **continent** the spinner stops at.
● Fill in the **databank** by putting an × in **one** of the squares for that continent.
● Fill in the **column grid** by shading **one** patch in the column for that continent. Use the **same colour** as that on the spinner.
● Shade **only one patch** for **each spin** until the correct number of patches in each column has been shaded.
● The winner is the first player to shade **all of the columns** and make a bar graph.

The spinner
Colour the sectors of the spinner. Follow the colour code in the table on page 11. For example, Europe is purple.

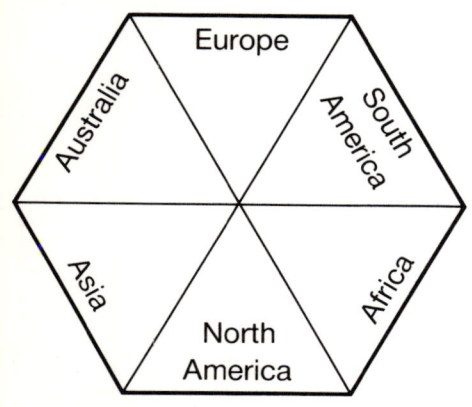

Databank
People per square km of cropland

North America
```
┌───┬───┬───┐
│   │   │   │
├───┼───┴───┘
│   │   50
└───┘
```

Africa
```
┌───┬───┬───┐
│   │   │   │
├───┼───┴───┘
│   │   50
└───┘
```

South America
```
┌───┬───┬───┬───┐
│   │   │   │   │
├───┼───┴───┴───┘
│   │   55
└───┘
```

Australia
```
┌───┐
│   │   5
└───┘
```

Asia
```
┌───┬───┬───┬───┬───┬───┐
│ × │   │   │   │   │   │
├───┼───┼───┼───┼───┼───┤
│   │   │   │   │   │   │
├───┼───┼───┼───┼───┼───┤
│   │   │   │   │   │   │
├───┼───┼───┼───┼───┼───┤
│   │   │   │   │   │   │
└───┴───┴───┴───┴───┴───┘
                        240
```

Europe
```
┌───┬───┬───┬───┬───┬───┐
│   │   │   │   │   │   │
├───┼───┼───┼───┼───┼───┤
│   │   │   │   │   │   │
├───┼───┼───┼───┼───┼───┤
│   │   │   │   │   │   │
├───┼───┼───┼───┼───┼───┤
│   │   │   │   │   │   │
└───┴───┴───┴───┴───┴───┘
                        210
```

The databank shows how many people there are for each square kilometre of cropland in each of the six continents.

For example, there are 210 people for every square kilometre of cropland in Europe.

Each square stands for 10 people. So, for Europe there are 21 squares.

Put an × in each square as the spinner stops. For example, if it stops on Asia, put an × in one of the Asia boxes.
One has been put in for you, giving you a free go.

The first patch in the Asia column on the graph on page 17 has also been shaded in.

Constructagraph *(continued)*

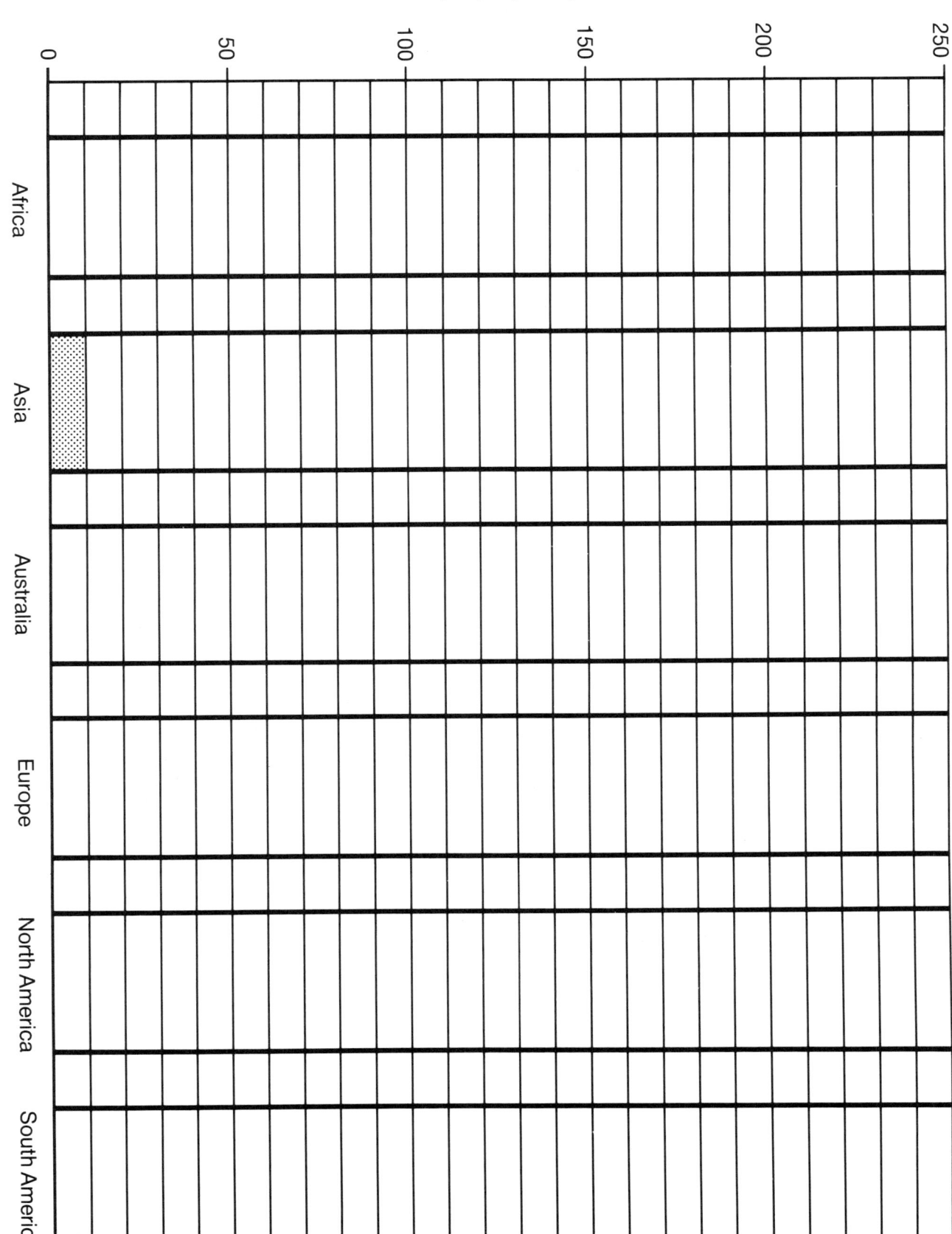

Play-a-Line

This is a game for two players.

How to play

- You each will need your **own grid**, a different coloured pencil and a ruler. A die is also needed. The grid is drawn below.
- Decide who throws the die first.
- Take turns to throw the die and plot on **your own grid** the number **you throw**. For example, if you throw 3 on your first throw, you mark × on the grid at A3. Your second throw will be marked at B, and so on.
- There is a shaded **penalty area** on the grid. If the number you throw falls into this area, you **miss a turn**.
- Each player must **record each throw** in the boxes below the grid.
- The winner is the first player to reach J.
- Draw a line by **joining up the ×**.

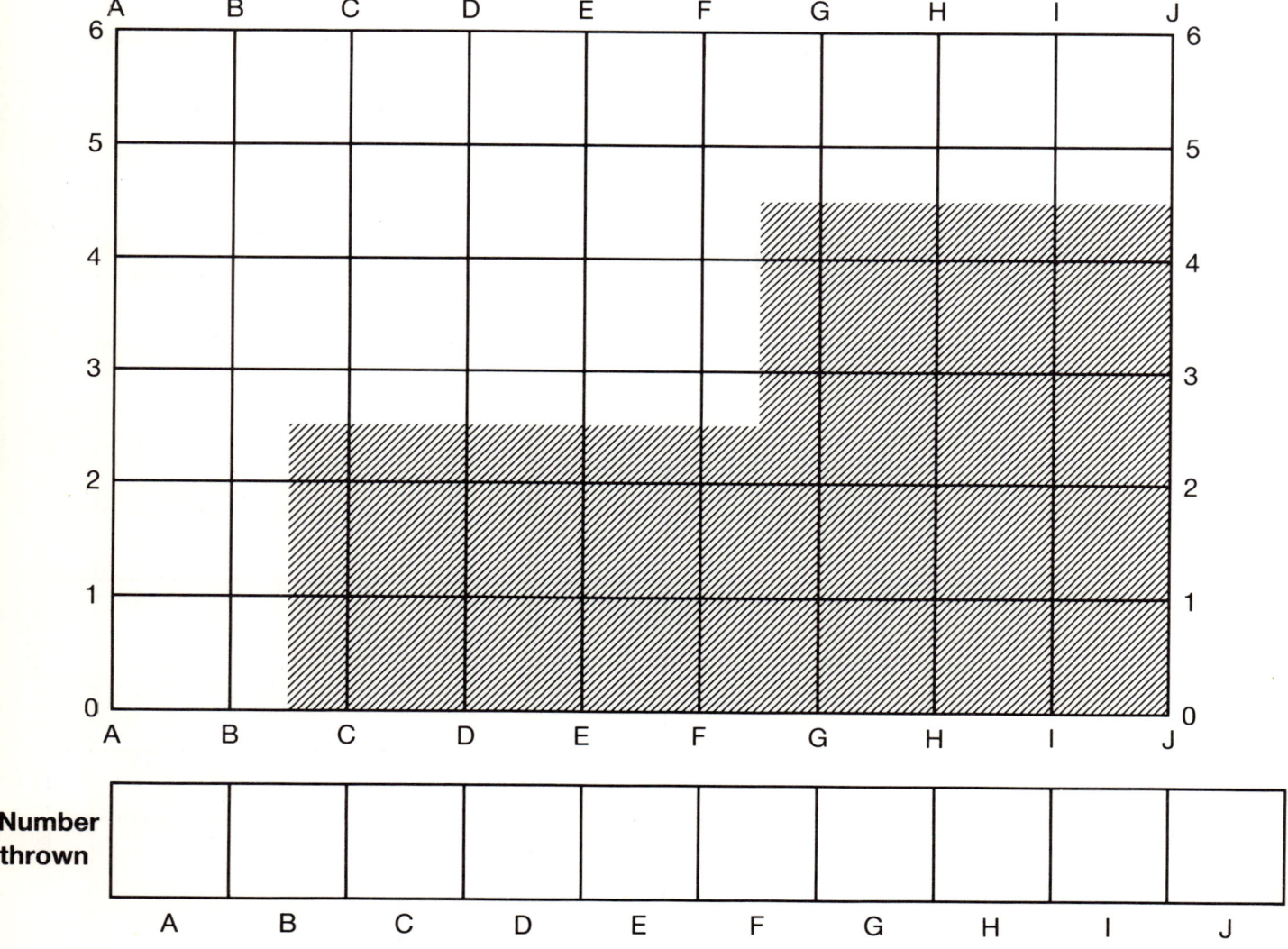

You have drawn a line graph.

Rising Star

This is a game for two players.

How to play

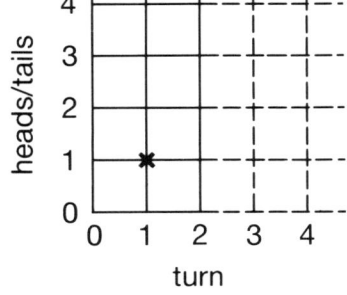

- Each player needs a copy of the grid drawn below. A coin is also needed.
- Decide who will toss **heads** and who will toss **tails**.
- The heads player starts by tossing the coin.
- If heads is tossed, the heads player marks × on his/her **own grid** where the **vertical line 1** (the 'Turn' line) crosses the **horizontal line 1** (the 'Heads or Tails' line).
- The diagram on the left shows this example.
- The tails player tosses second.
- If tails comes up, the tails player marks × on his/her grid.
- A player who tosses the **wrong side** of the coin, does **nothing**.
- The next time a player correctly tosses the coin, he/she marks × on the grid where **line 2** crosses **line 2**. On the next correct throw, an × is marked where **line 3** crosses **line 3**, and so on.
- The winner is the first player to reach 15.
- Join each ×. Your graph should be a **straight line** from the bottom left-hand corner of the grid to the top right-hand corner.

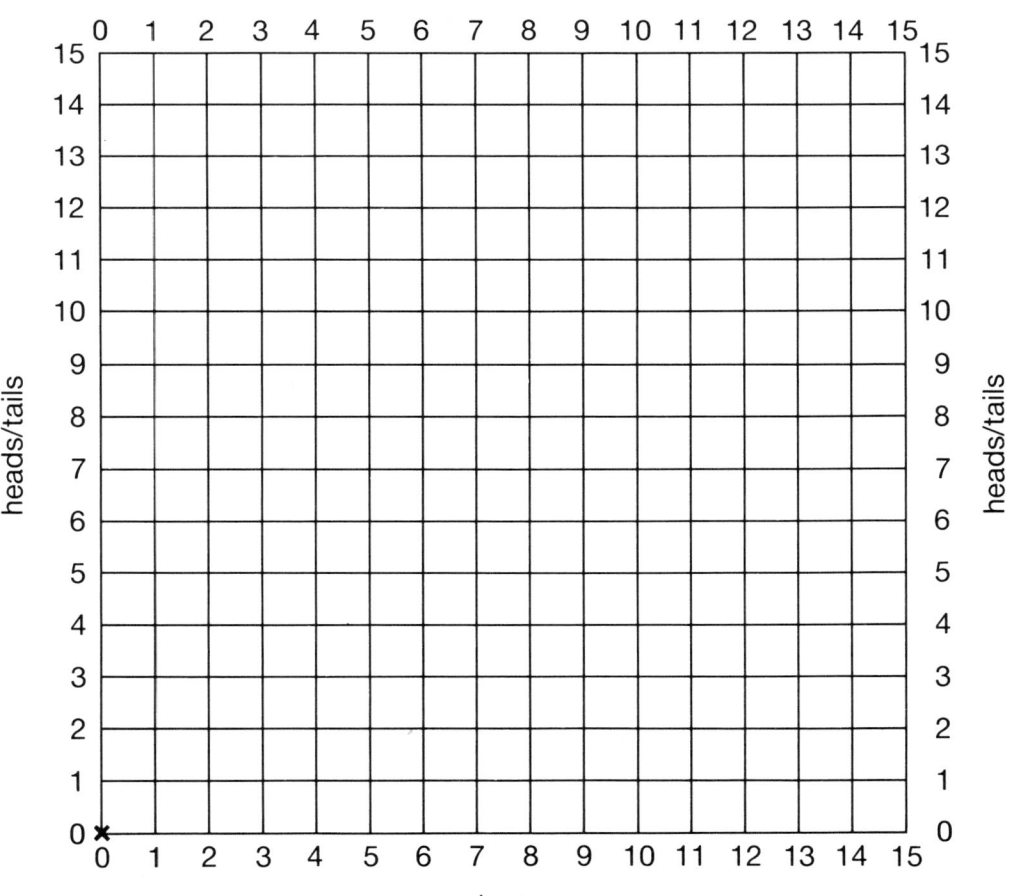

Temperature and line graphs

On the **bar graph** below, the height of each bar shows the **average temperature** in London for each month of the year.

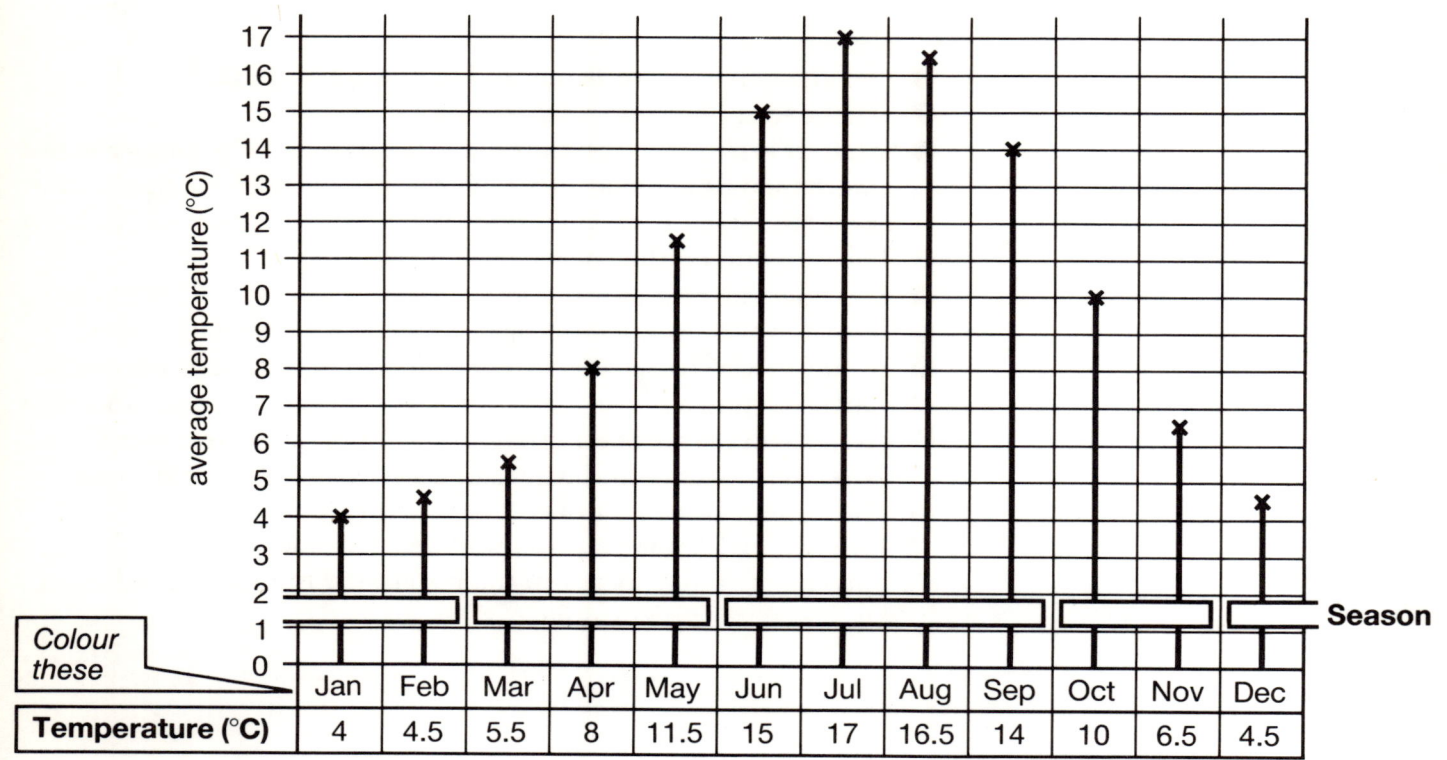

Temperature (°C)	Jan	Feb	Mar	Apr	May	Jun	Jul	Aug	Sep	Oct	Nov	Dec
	4	4.5	5.5	8	11.5	15	17	16.5	14	10	6.5	4.5

Colour these

Season

1 What is the **lowest** temperature?_____°C

2 Which **three months** have temperatures of 4.5 °C or less?

_____

3 What is the **highest** temperature?_____°C

4 Which **month** has the **highest** temperature?_____

5 What is the difference between the **highest** and **lowest** temperature?

    Circle 10 °C 13 °C 16 °C

From a bar graph like the one above, it is difficult to tell when the temperature begins to change more quickly. A **line graph** shows that more clearly.

Each bar on the temperature graph above has a × at its top. Join up the × using a ruler.

You have drawn a line graph.

Name:

Temperatures in London

Look again at the **line graph** you drew on page 20.

You will see that the names of the months are shown in the boxes below the graph.

What to do

● Use **red** to shade the names of the months which have an average temperature of **13 °C or more**.
● Use **red** to shade the columns above these months, up to the line you drew joining the ×.
● Use **blue** to shade the names of the months which have an average temperature of **5 °C or less**.
● Use **blue** to shade the columns above these months, up to the line you drew joining the ×.
● Use **green** to shade the names of **March**, **April** and **May**. Shade the columns also.
● Use **brown** to shade the names of **October** and **November**. Shade the columns also.

1 Which months are **spring**?

_____

2 Which months are **autumn**?

_____

3 Which months are **summer**?

_____

4 Which months are **winter**?

_____

5 Write **spring**, **autumn**, **summer**, **winter** in their correct places on the graph.

6 Complete the **title** of the map. In an atlas, find a map of Great Britain. Find **London** on the map. Mark London on the map on this page.

7 Use the atlas to find the latitude and longitude of London.

London is at **latitude** _____ and **longitude** _____

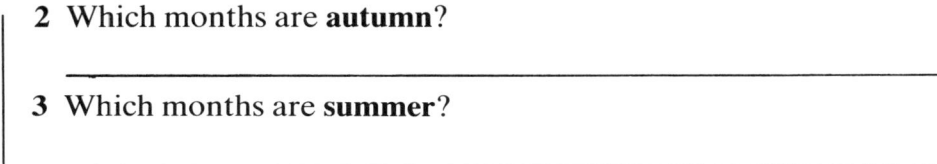

longitude 0°

N
W ——|—— E
S

A map of Great Britain showing _____

_____

latitude 50°N

Temperature chart for Athens

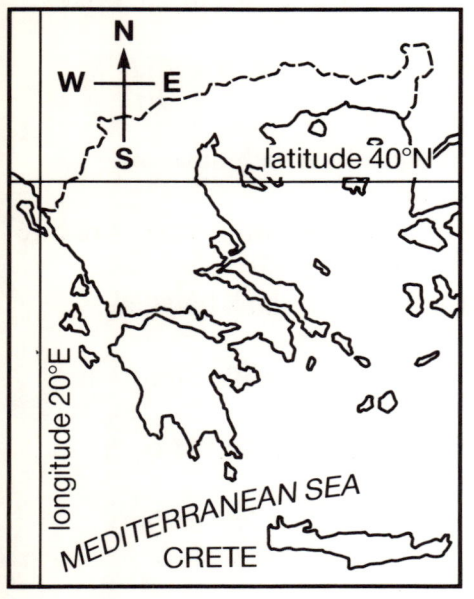

Here is a map of Greece.

What to do

● Find Greece in an atlas. Then find **Athens**.
● Mark Athens in the correct place on the map on this page.
● Use the **grid** below and the **temperatures** in the **databank** at the bottom of this page to plot a **line graph** for Athens.

The **vertical axis** shows **temperatures** from 0 °C to 30 °C.
The **horizontal axis** shows the **months** of the year from January to December.

● Write the names of the **months** in the spaces below the grid. The first letter of each is given. January has been done for you.

● Mark **January's temperature** (9 °C) with ×, where the 9 °C temperature line crosses the month line. February, July and December have been done for you.

● Show the **temperatures** of the remaining months by marking × at the **correct heights** on the month lines.

● Complete the graph by joining up the ×. Use a ruler to draw your line.

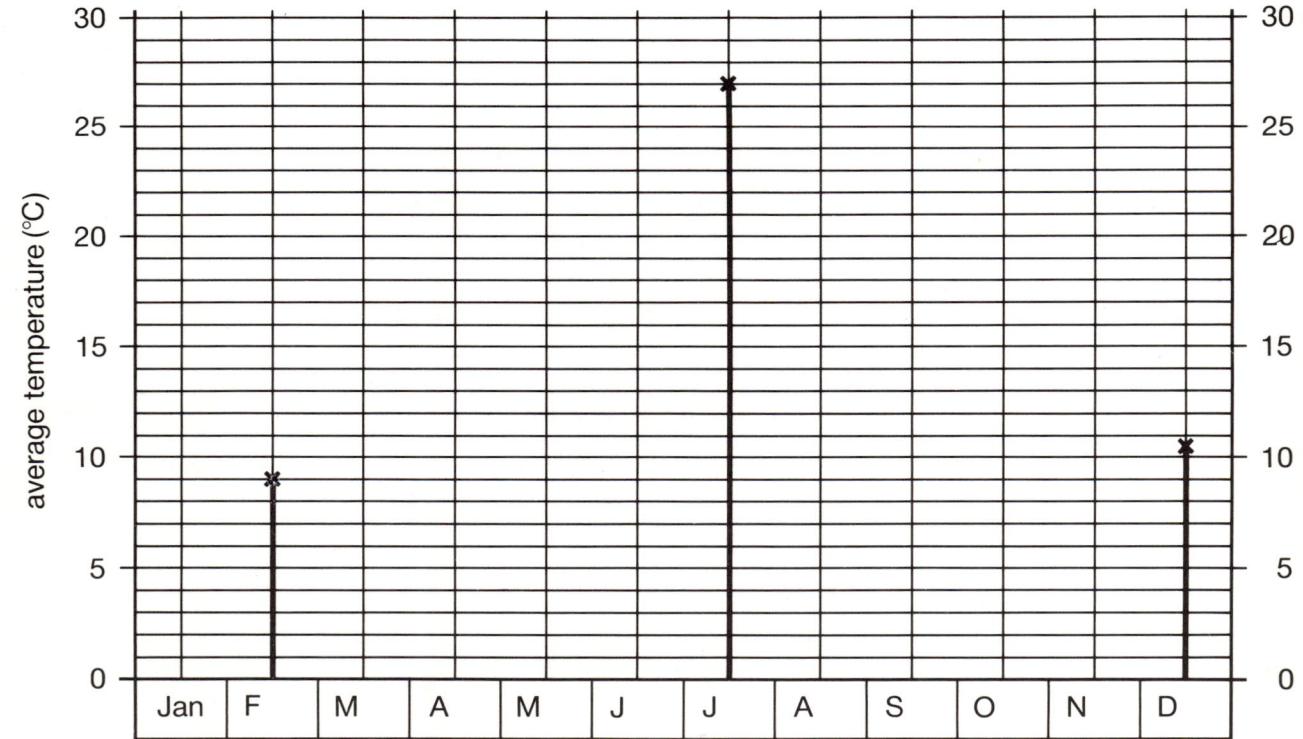

Databank												
	Jan	F	M	A	M	J	J	A	S	O	N	D
Average temperature (°C)	9	9	11.5	15	20	24.5	27	27	23	19.5	14.5	10.5

Two temperature graphs on one grid: Athens and London

The information in the **temperature databank** below left is the same as the information you have used in the graphs on pages 20 and 22.

When both graphs are plotted on one grid, it is much easier to compare them.

Databank		
Temperatures (°C)		
Month	Athens	London
Jan	9	4
Feb	9	4.5
Mar	11.5	5.5
Apr	15	8
May	20	11.5
Jun	24.5	15
Jul	27	17
Aug	27	16.5
Sep	23	14
Oct	19.5	10
Nov	14.5	6.5
Dec	10.5	4.5

What to do

● At the side and along the bottom of the grid, name the **vertical axis** and the **horizontal axis**. The horizontal axis marks the months of the year. The vertical axis marks the temperatures in degrees Celsius (°C).

● **Plot the temperatures for London**. The January temperature is 4 °C. It has been plotted for you with a ×. Plot the other 11 temperatures, one for each month, also with ×. Join the × with a **blue** line.

This is the temperature graph for London. Name it on the line you have drawn.

● **Plot the temperatures for Athens**. The January temperature is 9 °C. It has been plotted for you with a **dot**. Plot the other 11 temperatures, one for each month, with a **dot**. Join the **dots** with a **red** line.

This is the temperature graph for Athens. Name it on the line you have drawn.

● Using a **yellow** pencil, shade the space between the two lines. This shows the **difference** between the two temperature graphs.

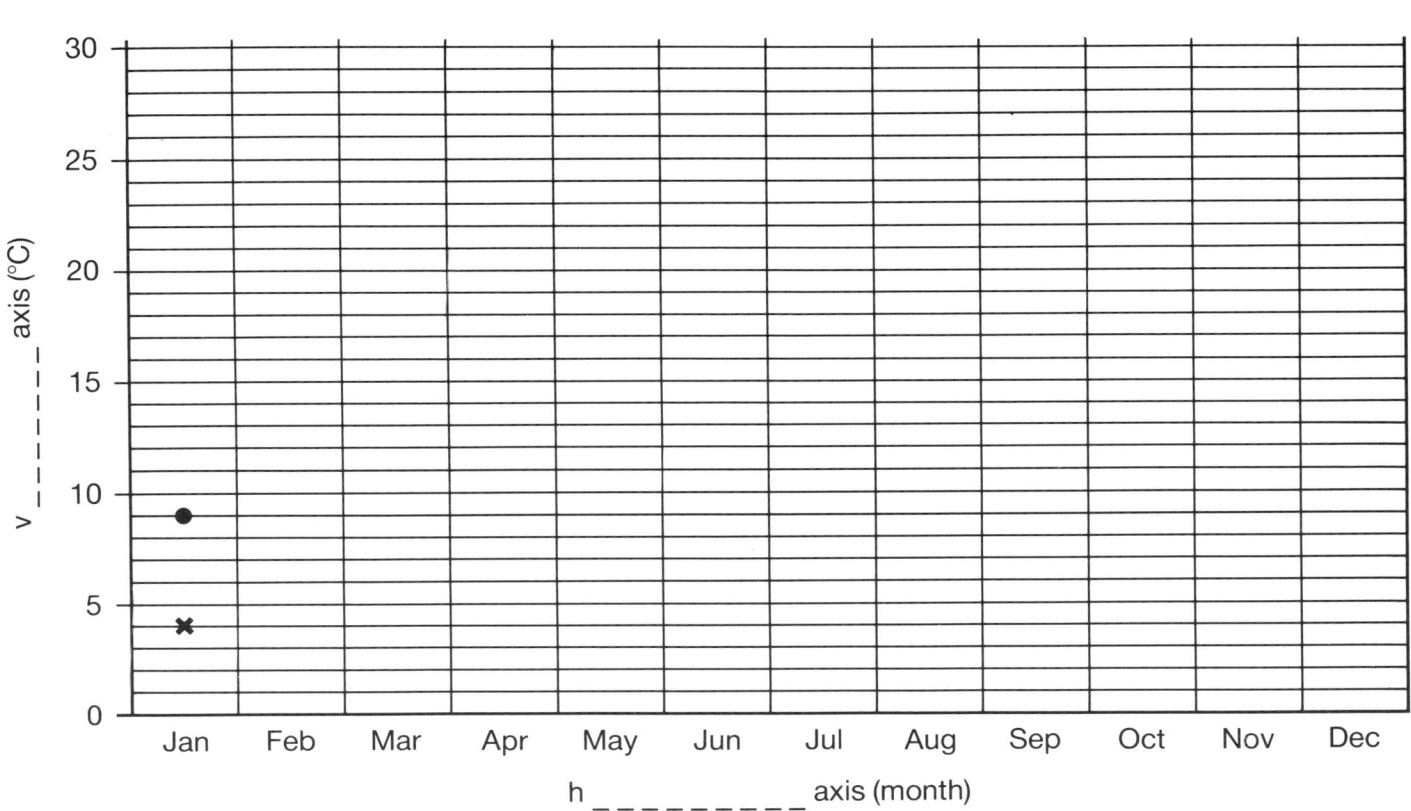

Two graphs on one grid (continued)

Look again at the graph on the previous page.

1 Which city is the **warmer**, Athens or London? _____

2 What is the **highest temperature** for Athens? _____ °C

3 What is the **highest temperature** for London? _____ °C

4 What is the **lowest temperature** for Athens? _____ °C

5 What is the **lowest temperature** for London? _____ °C

6 Which **month** has the **highest temperature**
   for both cities? _____

7 Which **month** has the **lowest temperature**
   for both cities? _____

8 How much **warmer** is Athens than London in **July**? _____ °C

9 How much **warmer** is Athens than London in **January**? _____ °C

10 Athens is **always warmer** than London.   *Circle* TRUE/FALSE

11 The **difference in temperature** between
    London and Athens is the same in July
    as it is in January.   *Circle* TRUE/FALSE

Temperature range

12 The **temperature** in Athens in July is _____ °C

13 The **temperature** in Athens in January is _____ °C

14 **The difference** between July and January is _____ °C

The **difference** between the **highest** and the **lowest** temperatures is called the **temperature range**.

15 The **temperature range** for London:

July temperature _____ °C

January temperature _____ °C

Temperature range _____ °C

4. LINE AND BAR GRAPHS   Name:

Temperature difference between London and Athens

Athens is warmer than London.

A bar graph can be used to show the **difference** between the two.

Databank													
Month													
	Jan	Feb	Mar	Apr	May	Jun	Jul	Aug	Sep	Oct	Nov	Dec
Temperature (°C)													
Athens	9												
London	4												
Difference	5												

What to do in the databank

- ● Write in the **average monthly temperatures** for Athens. January has been done for you.
- ● Write in the **average monthly temperatures** for London. January has been done for you.
- ● Take away the London temperature from the Athens temperature for each month. Write the answers in the **difference** boxes. January has been done for you.

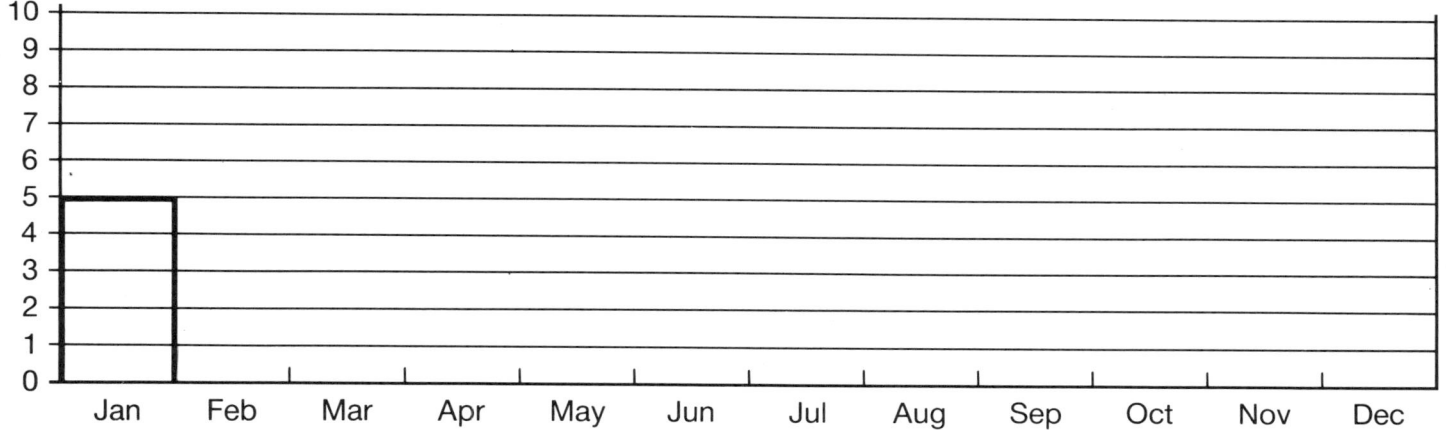

What to do on the grid

- ● Complete the bar graph to show the **difference** in temperature between London and Athens.

The difference in temperature for January is _____°C. A bar has been drawn to the correct height for you.

- ● Draw a bar for each of the other months. The height of each bar is the **temperature difference in** °C which you worked out in the **difference boxes.**
- ● Using **blue**, colour the bars which equal a temperature difference of **8.5 °C or less**.
- ● Using **red**, colour the bars which equal a temperature difference of more than **8.5 °C**.

The red bars show the months when the difference between the temperatures in the two cities is greatest.

Hot and Cold

Look at the work on pages 20 – 24. Then complete the crossword. The clues for this crossword are below.

Across

   1 The month with average temperatures of 10.5 °C and 4.5 °C.
   4 The month with average temperatures of 20 °C and 11.5 °C.
   5 The month with average temperatures of 11.5 °C and 5.5 °C.
   9 The month with average temperatures of 23 °C and 14 °C.
  10 The month with average temperatures of 14.5 °C and 6.5 °C.
  11 The city near 38°N 24°E.
  12 The month with an average temperature of 27 °C and 16.5 °C.
  13 The month with an average temperature of 27 °C and 17 °C.
  14 The city near 50°N and the Greenwich Meridian.

Down

   2 The month with average temperatures of 24.5 °C and 15 °C.
   3 The month with average temperatures of 19.5 °C and 10 °C.
   6 The month with average temperatures of 15 °C and 8 °C.
   7 The month with average temperatures of 9 °C and 4.5 °C.
   8 The month with average temperatures of 9 °C and 4 °C.

Have a Good Holiday

This is a game for three or four players.

The aim is to try to get the best July weather you can at an English seaside resort. This means you want as many warm, dry days as possible.

How to play

● You will need a pack of playing cards. Each card shows the weather for **one day of your holiday**. The meaning of each card is shown on the weather table on page 28.
● Decide who will deal and who will play first.
● The dealer deals **seven** cards to each player. The remaining cards (the pack) are put **face down** in the middle of the table.
● Each player picks up his/her hand of cards and looks at them to see what the weather will be.
● The first player takes the top card from the

pack. If this player wants to keep this card, then he/she puts down **another** card **face up** next to the pack.
● The second player can choose **either** the top card from the pack **or** the card which is face up. If this player wants to keep either card, then he/she must put down **another** card, **face up** next to the pack. And so on . . .
● Each player, in turn, must choose and put down a card. However, if a player has only good-weather cards in his/her hand, then that player can simply say **pass** and **not pick up** or put down a card.
● The highest cards have the best weather. However, if you collect **seven** cards of the **same suit** (for example, seven diamond cards), these will count as **seven aces**.
● The game continues until there are **no** cards in the pack. When this part of the game is finished, you should have **seven cards in your hand**. Use these cards to complete the **Holiday Weather Databank** below. You can put the weather in any order you wish.

Holiday weather databank							
	Sunday	**Monday**	**Tuesday**	**Wednesday**	**Thursday**	**Friday**	**Saturday**
Card							
Temperature (°C)							
Hours of rainfall							
Hours of sunshine							

On the grids on page 28 draw a **line graph** for **temperature** and a **bar graph** for **rainfall**.

Look at the graphs you have drawn and write six lines explaining whether you have had a good holiday or a bad holiday.

_____

_____

_____

_____

_____

Have a Good Holiday
(continued)

Line graph showing highest daily temperatures

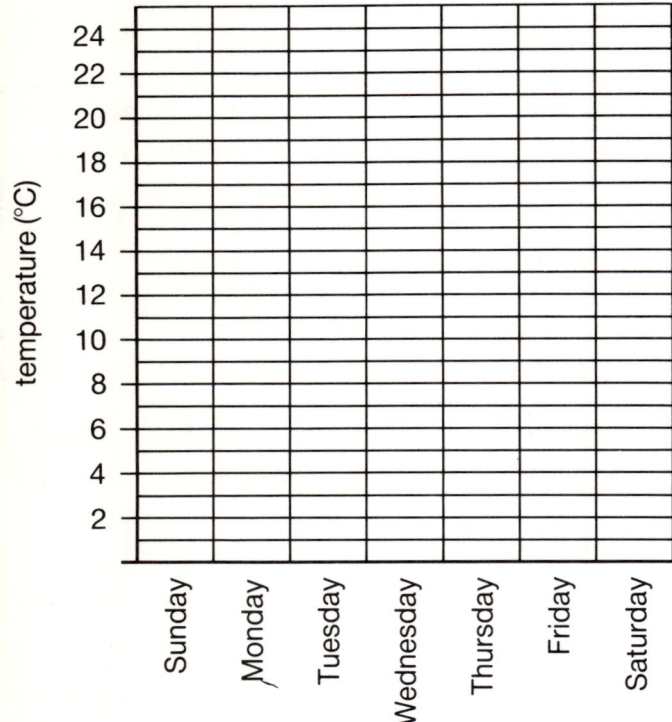

Bar graph showing hours of rainfall

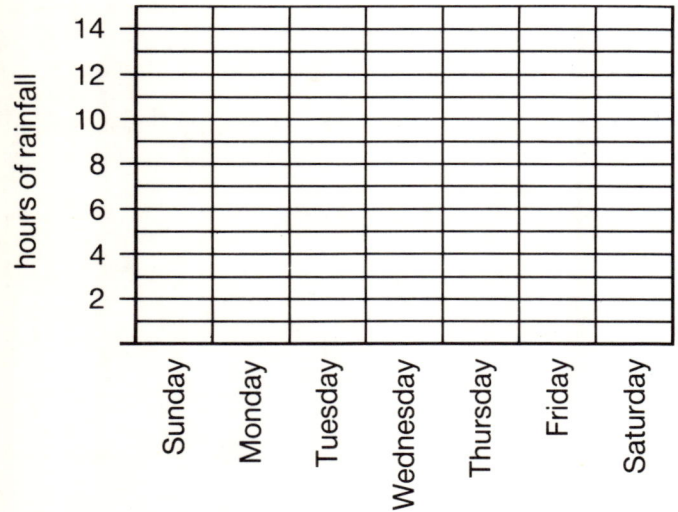

Weather table

ACE
    24 °C highest temperature
    14 hours sunshine
     0 hours rainfall during daylight

KING
    23 °C highest temperature
    13 hours sunshine
     0 hours rainfall during daylight

QUEEN
    22 °C highest temperature
    12 hours sunshine
    ½ hour rainfall during daylight

JACK
    21 °C highest temperature
     8 hours sunshine
     3 hours rainfall during daylight

TEN
    20 °C highest temperature
     8 hours sunshine
     1 hour rainfall during daylight

NINE
    19 °C highest temperature
     9 hours sunshine
     2 hours rainfall during daylight

EIGHT
    18 °C highest temperature
     9 hours sunshine
     4 hours rainfall during daylight

SEVEN
    17 °C highest temperature
     7 hours sunshine
     0 hours rainfall during daylight

SIX
    16 °C highest temperature
     1 hour sunshine
     0 hours rainfall during daylight

FIVE
    15 °C highest temperature
     2 hours sunshine
     6 hours rainfall during daylight

FOUR
    14 °C highest temperature
     2 hours sunshine
     8 hours rainfall during daylight

THREE
    14 °C highest temperature
     1 hour sunshine
     8 hours rainfall during daylight

TWO
    13 °C highest temperature
     0 hours sunshine
    14 hours rainfall during daylight

Package Holiday

This is a game for three or four players.

How to play

● This game is played the same way as Have a Good Holiday except that you have to choose one of the following types of holiday before you deal the cards.
● Look at the holidays below. Write your choice of holiday in the space above the databank.

Holiday type	Activities	Weather
Outdoor in south-west England	Hiking, canoeing, cycling, pony riding, sailing, swimming	Dry, not too hot. Ideally 16 to 19 °C
In London	Visiting museums, shops, art galleries, theatres, sightseeing	Not too hot, ideally not above 20 °C, rain only really affects sightseeing
In Blackpool	Sunbathing, boat trips, fun fair, amusement arcades, variety shows, circus	Hot, dry weather for outdoor activities, but the weather does not matter for other activities

● When you have chosen your holiday and written down your choice, play the game using the same rules as those for Have a Good Holiday.
● When there are no cards left in the pack, fill in the databank below. You should also write what you did in the 'Activity' boxes.

My choice of holiday: _____

Holiday databank							
	Sunday	**Monday**	**Tuesday**	**Wednesday**	**Thursday**	**Friday**	**Saturday**
Card							
Temperature (°C)							
Hours of rainfall							
Hours of sunshine							
Activity							

Did you have a good holiday? _____

**Top Secret
Temperature Code**

On page 31 is a coded message. To decode it, you must
- Make a code breaker.
- Get the number clues 1 to 14.
- Fill in the security ribbon which is on page 31.

Making the code breaker

The code breaker is on page 32.
- In the left-hand column of the code breaker, write the 26 letters of the alphabet in order. A and Z have been done for you.
- In the right-hand column, write the numbers 1 to 26. 1 and 26 have been done for you.

Getting the number clues

You will get the answers to questions 1 to 14 below by reading the **code breaker graphs** on page 32 next to the code breaker. The map on that page shows where London, Stavanger, Lisbon and Lagos are.
- The answers to these questions are temperatures in °C.
- Write the answer to each question in the box on the left next to that question number.

1	
2	
3	
4	
5	
6	
7	
8	
9	
10	
11	
12	
13	
14	

1 What is the temperature of Stavanger in January and February?

2 What is the temperature of London in February and December?

3 What is the temperature of London in November?

4 What is the temperature of London in May?

5 What is the temperature of Stavanger in June and September?

6 What is the temperature of the two hottest months in Stavanger?

7 What is the temperature of London in June?

8 What is the temperature of London in August?

9 What is the temperature of London in July?

10 What is the temperature of Lisbon in October?

11 What is the temperature of Lisbon in June?

12 What is the temperature of Lisbon in September?

13 What is the temperature of Lisbon in July?

14 What is the temperature of the coolest month in Lagos?

Top Secret Temperature Code *(continued)*

The security ribbon
Look at the security ribbon. The top row contains question numbers.

Question	11	2	5	7	2	9	1	11	12	9	2	10	1	9	2	3	6	11	11	2	9
Answer																8					
Letter																H					

Question	1	10	14	6	12	11	9	1	13	2	4	11	6	11	3	2	2	8	12	1	11	6	9
Answer															8								
Letter															H								

Write your answers to the questions on page 30 in the boxes in the middle of the ribbon. Each answer must be in the box under its question number. The answer to question 3 has been written in for you.

Now break the code

When you have written the answers in their boxes, you can break the code.

Each number in the middle row of the ribbon stands for a letter. The code breaker which you made on page 32 tells you the numbers and their letters. For example, number 1 equals A.

In the example which has been filled in on the security ribbon, question 3 has the answer **8**. So **8** has been written in.

According to the code breaker, **8** equals **H**. So **H** has also been written in the bottom row of boxes on the ribbon.

Wherever question 3 is shown in the top row, 8 is put into the middle row and H into the bottom row.

Carry on filling in the letters in the bottom row of the ribbon.

When you have decoded the message, look at the map on page 32.

Is the message **true** or **false**?   *Circle*   TRUE/FALSE

Top Secret Temperature Code (continued)

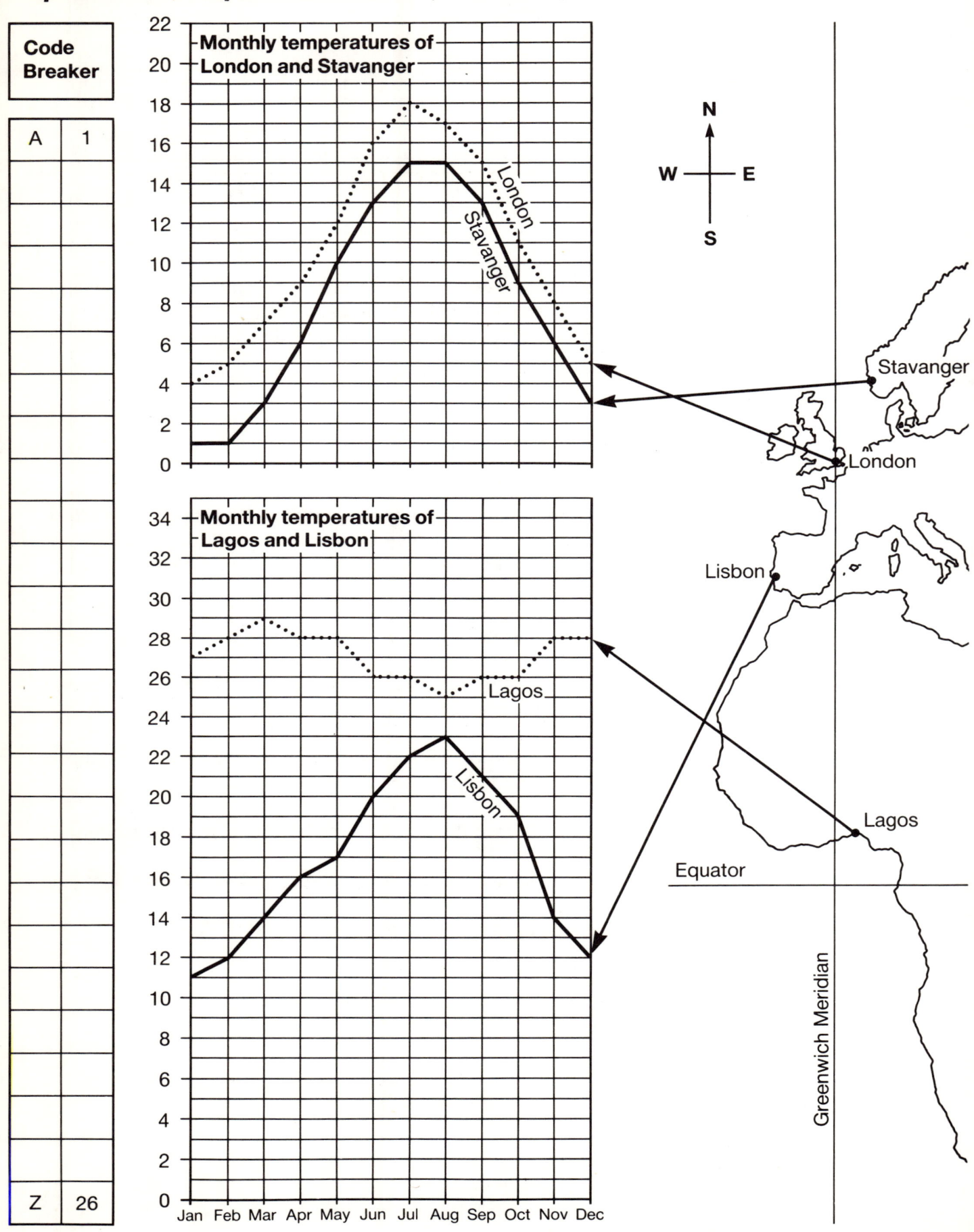

Code Breaker	
A	1
Z	26

Monthly temperatures of London and Stavanger

London
Stavanger

N
W — E
S

Stavanger

London

Monthly temperatures of Lagos and Lisbon

Lagos

Lisbon

Lisbon

Lagos

Equator

Greenwich Meridian

Jan Feb Mar Apr May Jun Jul Aug Sep Oct Nov Dec

Round the Pie Chart

This is a game for two or more players.

How to play

- You will need a die.
- You each have a copy of the divided circle on this page.
- Decide who will start the game.
- Take it in turn to throw the die.
- Shade **your own score** on **your own circle**. For example, if you throw **4**, shade **4 sectors**.
- Shade the score from each throw in a **different colour**.
- The winner of the game is the first player to complete the circle.
- The **exact number** must be thrown to complete the circle.

Start here and go round clockwise, shading each sector in turn.

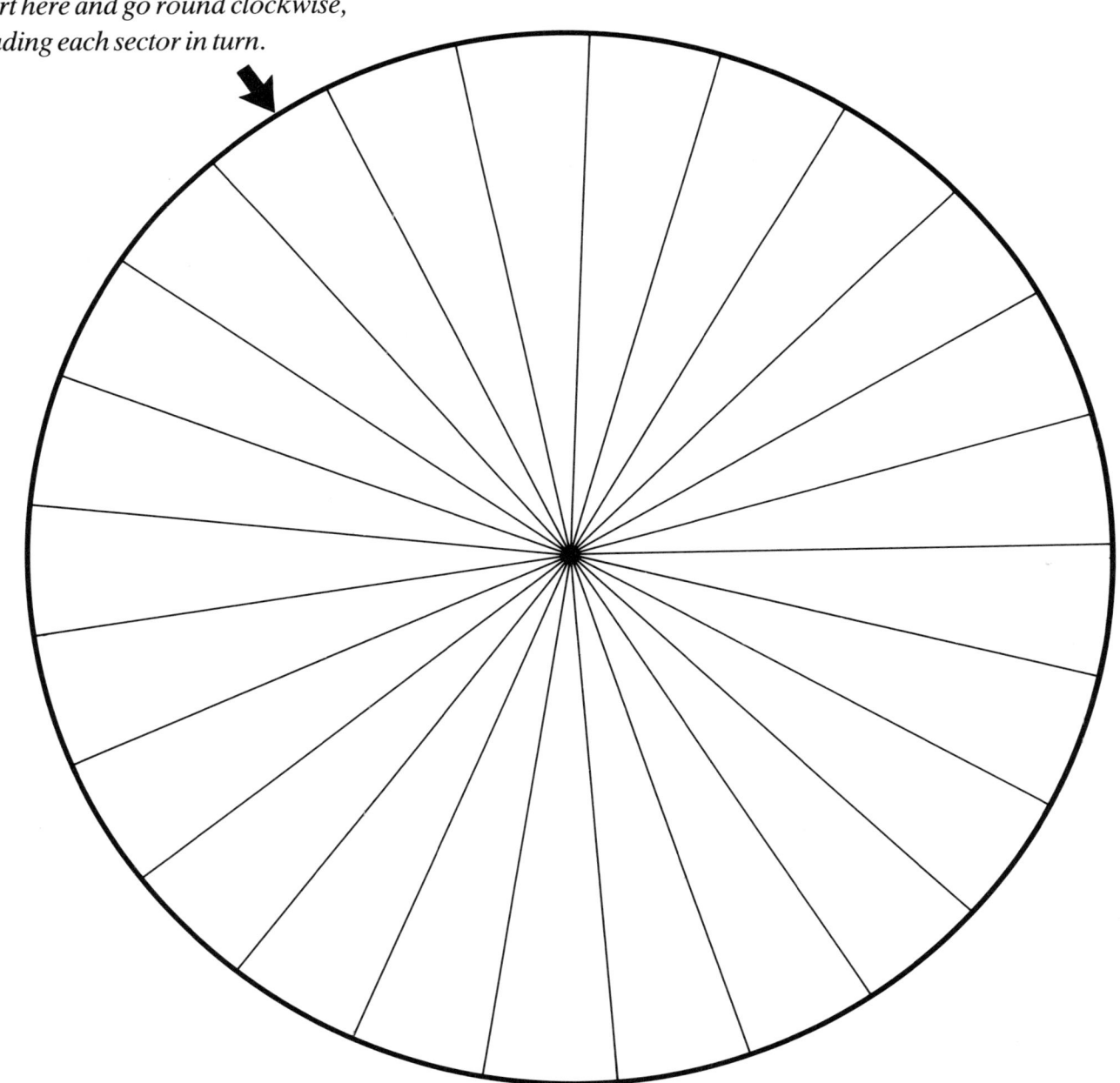

Where did people stay?

Twenty-five people were asked where they stayed on holiday last year. Their answers are given in the databank.

Where-they-stayed databank			
Hotels and boarding houses	With friends	Holiday camps	Self-catering accommodation
8	6	3	8

The circle below is divided into sectors which are numbered from 1 to 25. Each sector stands for one person.

What to do

● Use **red** to shade the number of people staying in **hotels and boarding houses**.
● Use **green** to shade the number of people staying with **friends**.
● Use **blue** to shade the number of people staying in **holiday camps**.
● Use **yellow** to shade the number of people staying in **self-catering accommodation**.
● Now complete the key by shading in the boxes and writing on the line next to each box what the colour means.

You have made a pie chart.

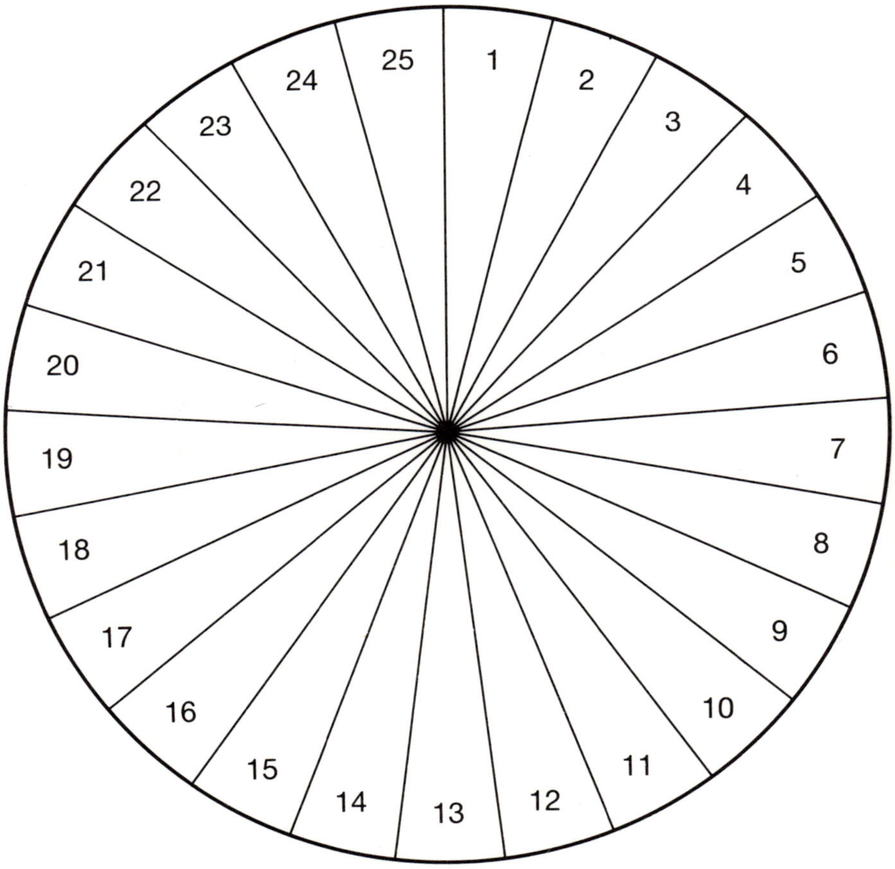

Key

[ ] _____

[ ] _____

[ ] _____

[ ]

With this pie chart you can easily see the **proportion** of people who stayed in each type of accommodation.

Name:

What do people want?

Twenty-five people were asked what they hoped to get from their holidays. Their answers are given in the databank.

Holiday-hopes databank				
Good food	Scenery	Good weather	Entertainment	Other things
2	3	9	8	3

What to do in the databank

- Shade the **good food** part **red**.
- Shade the **scenery** part **green**.
- Shade the **good weather** part **blue**.
- Shade the **entertainment** part **yellow**.
- Shade the **other things** part **brown**.

On the left and below are the sectors of a pie chart. Together they make up a whole pie chart.

Each sector stands for one of the answers in the databank and the number of people who gave it.

Each sector is divided to show how many people it stands for.

What to do with the sectors

With the help of the databank
- Shade the **good food** sector **red**.
- Shade the **scenery** sector **green**.
- Shade the **good weather** sector **blue**.
- Shade the **entertainment** sector **yellow**.
- Shade the **other things** sector **brown**.
- Then cut out the sectors and see page 36.

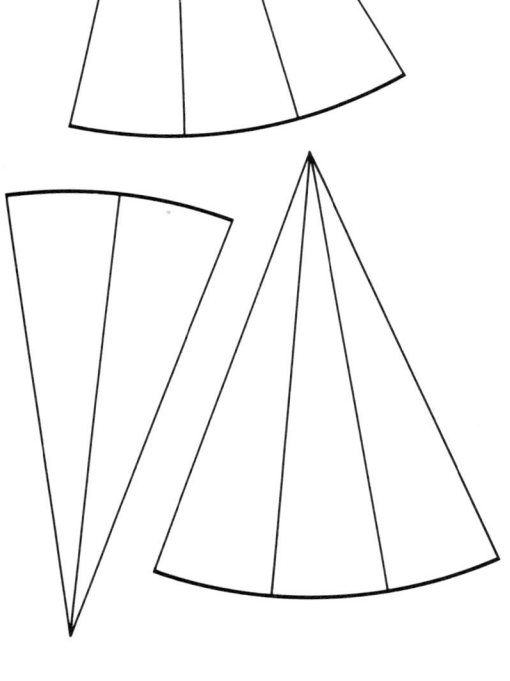

What do people want?
(continued)

Key

What to do

● Complete the pie chart by sticking down the coloured sectors which you have cut out.
● Complete the key by shading the boxes and writing in what each colour means.

1 What did **most people** hope for when they went on holiday?

_____

2 Which **two** reasons for going on holiday together make up **more than half** of the total number of people?

_____

3 What did the **smallest** group of people want?

_____

4 Which would have been **most important** to you?

_____

● Make a new databank showing what the **people in your class** would hope to have when they go on holiday this year.
● Now use your new databank to make a new pie chart.

Going to Spain

Twenty-five people went to Spain for their holidays. The databank shows when they went.

	Jan–Mar	Apr–Jun	Jul–Sep	Oct–Dec	Total
Number of people	3	6	12	4	25

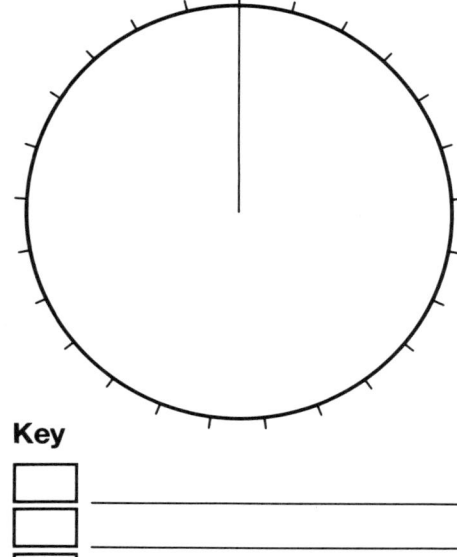

Key

● Use the circle given on the left to make a pie chart showing the information given in the databank. Also complete the key.

The pie chart shows that:

1 Most of the people went to Spain during the months of

_____

2 Fewest of them went there during the months of

_____

3 How many people went to Spain between **April** and **September**?

_____

Below is a map of Spain and the names of some well-known places there.

● Use an atlas to mark these places in their correct position on the map.

The places are **Madrid**, **Barcelona**, **Alicante**, **Malaga**.

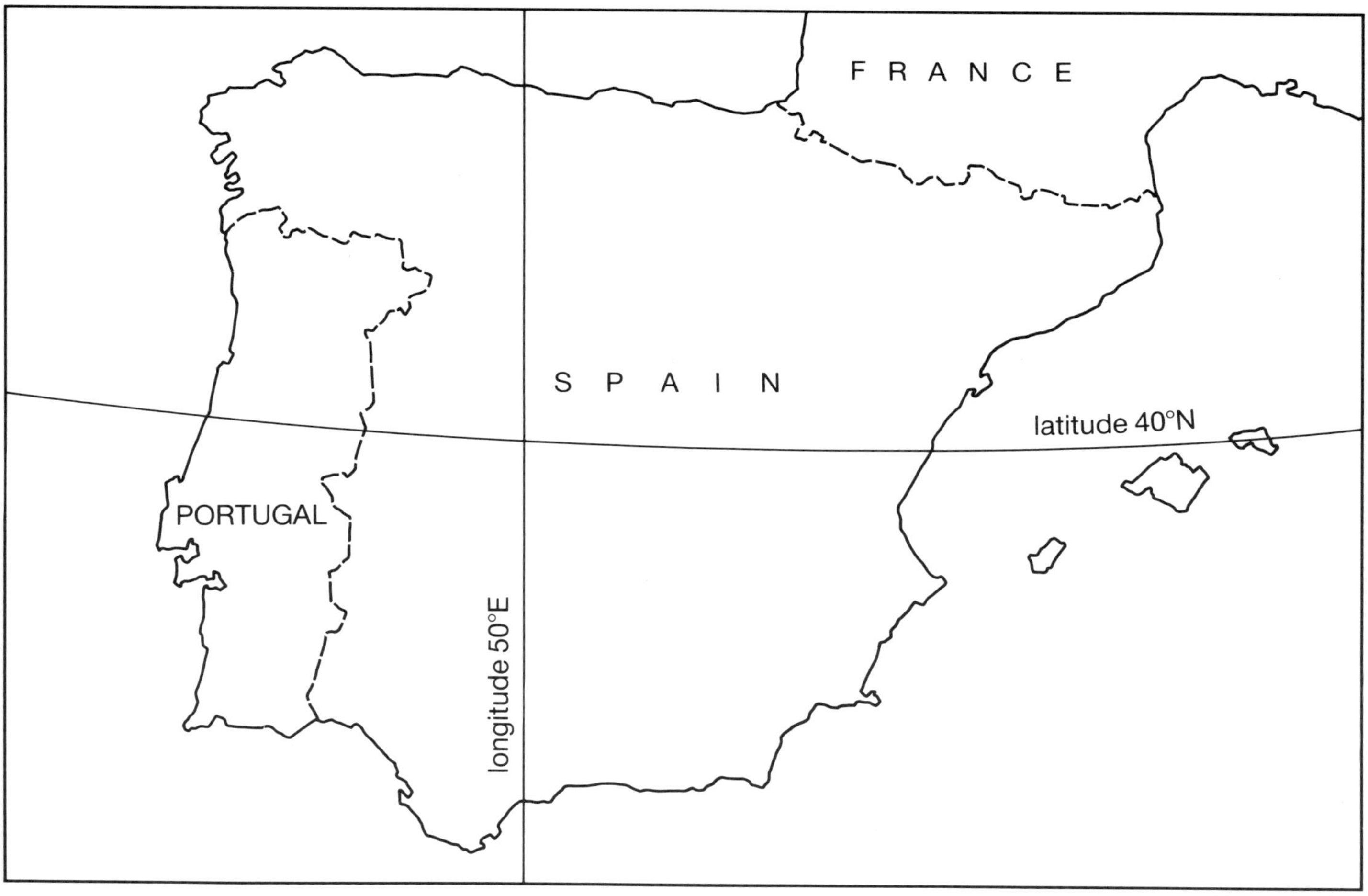

FRANCE

SPAIN

latitude 40°N

PORTUGAL

longitude 50°E

Why go to Spain in summer?

What to do

● Use the information in the databank to make a **line graph** on the grid to show **average monthly temperatures** in Malaga.

● Label the **horizontal** and **vertical axes**. Write in the boxes.

● The first temperature has been marked for you. Note that only the first letter of each month is given.

● When you have done all this, use the information in the databank to make a **bar graph** to show the rainfall in Malaga. The bar for January has been drawn for you.

Line graph

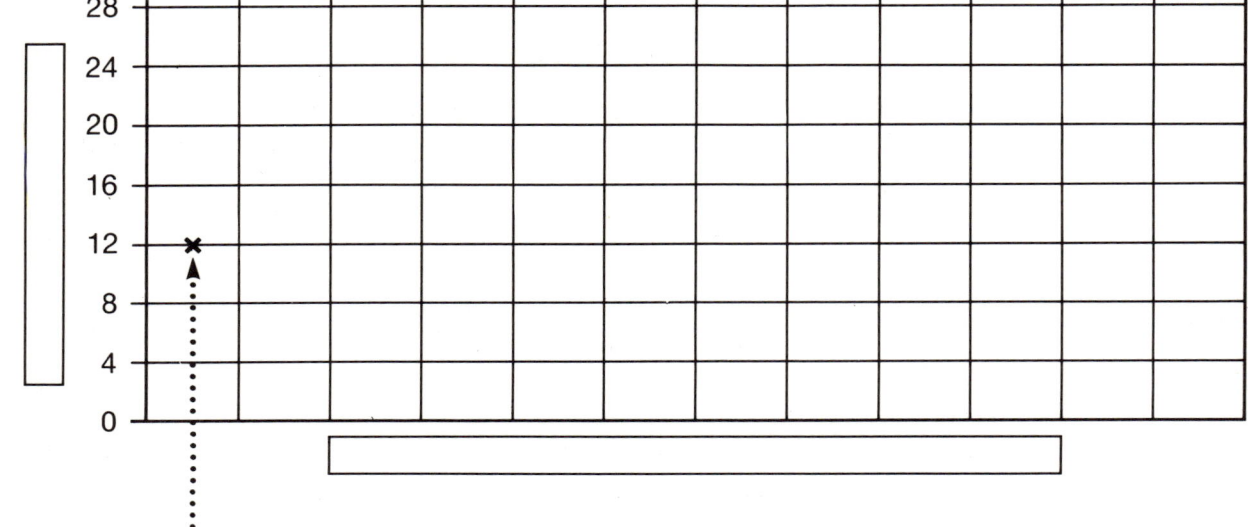

Databank: Malaga

Month	J	F	M	A	M	J	J	A	S	O	N	D
Temperature (°C)	12	13	15	17	19	23	25	26	23	20	16	13
Rainfall (mm)	61	51	62	46	26	5	1	3	29	64	64	62

Bar graph

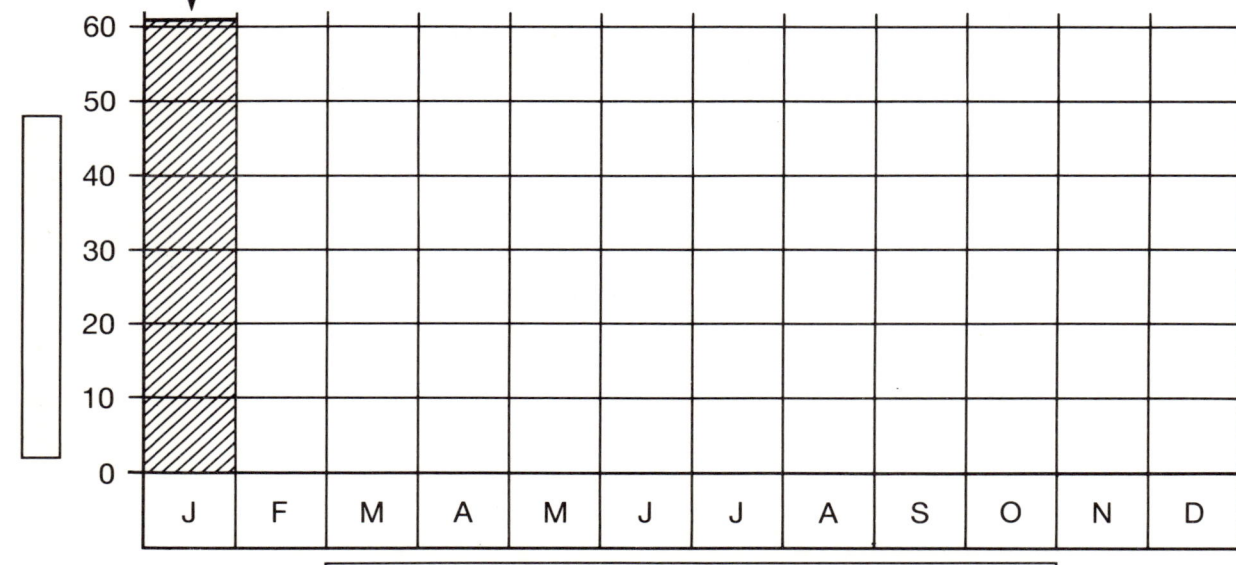

Why go to Spain in summer? *(continued)*

What to do

Fill in the missing words. They are given in the bottom box.

The four _____ months on the Mediterranean coast of Spain are June, July, August and September. The hottest month is _____ , when the temperature averages _____ , but _____ is only one degree cooler.

These two months also have _____ rainfall. July has _____ and August has _____ .
From October to _____ rainfall is much _____ and the temperatures are generally _____ than during the summer months. Therefore, people who want to go to Spain for warm, dry _____ will go during _____ .

In the sample of 25 people, more went to Spain for a holiday during _____ than during any other period of the year. _____ people went during the month of January, February and March, but not many went during _____ .

The missing words are:

July, August and September	July
fewest	least
October, November and December	1 mm
hottest	3 mm
March	26 °C
lower	August
June, July and August	heavier
weather	

Name:

In time for United

Imagine you live in Weston. Here is a simplified map of your town.

Your home and West Stadium are shown on it. West Stadium is the home ground of Weston United Football Club.

1 Is your home **north** or **south** of the town centre? (Look at the compass directions at the top right-hand corner of the map.)

2 What is the **grid reference** of your home? _____

3 Is the stadium **north-east** or **south-west** of your home? _____

4 What is the **grid reference** of the stadium? _____

Three bus routes cover the streets shown on the map. Bus 62 runs from **1**(home), through **2**(town centre) to **4**(East Side).

What to do

● Mark the route of **bus 62** in **red** on the map.
● Complete the sentences in the box on the left which describe the bus route.

The bus route begins at

my h _ _ _ and goes

s _ _ _ _ _ to the

t _ _ _ c _ _ _ _ _ —.

It then goes e _ _ _ to

E _ _ _ S _ _ _.

Missing words:
Side   east   home   East
town   south   centre

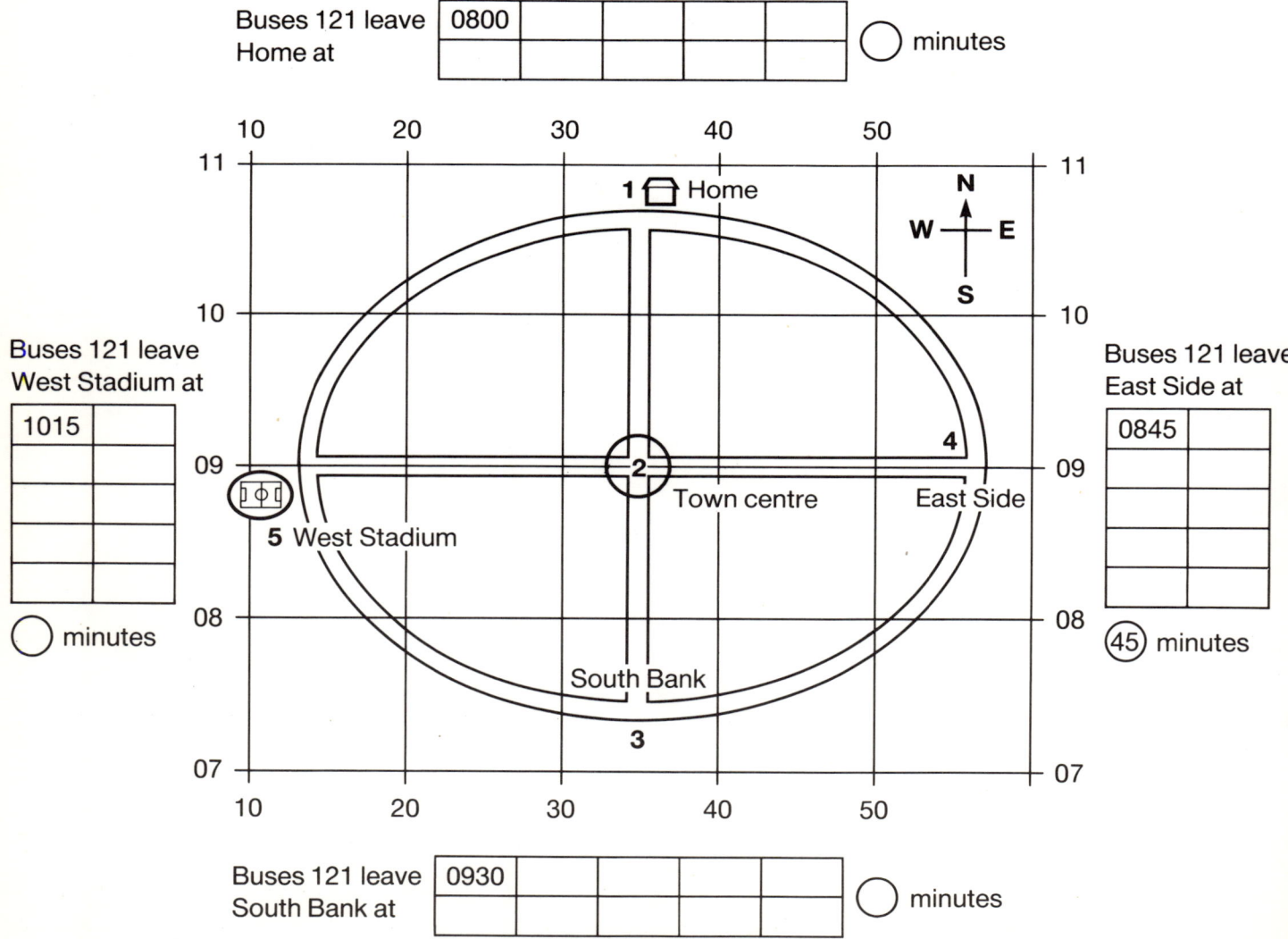

Buses 121 leave Home at | 0800 | | | | |

○ minutes

Buses 121 leave West Stadium at

1015	

○ minutes

Buses 121 leave East Side at

0845	

㊺ minutes

Buses 121 leave South Bank at | 0930 | | | | |

○ minutes

In time for United
(continued)

Here is the timetable for bus 62.

Stops	Journeys						
	A	B	C	D	E	F	G
1 Home *Richmond*	0700	0900	1100	1300	1500	1700	1900
2 Town centre	0725	0925	1125	1325	1525	1725	1925
4 East Side	0750	0950	1150	1350	1550	1750	1950
4 East Side	0800	1000	1200	1400	1600	1800	2000
2 Town centre	0825	1025	1225	1425	1625	1825	2025
1 Home	0850	1050	1250	1450	1650	1850	2050

Timetable

From the timetable, you can tell that the first bus of the day (A) leaves stop 1 at 7 am. It takes 25 minutes to get to the town centre, arriving there at 7.25 am. It takes another 25 minutes to get to East Side, arriving there at 7.50 am. The driver has a 10 minute break. The bus then returns from East Side to stop 1 by way of the town centre.

1 **How long** does it take a bus to travel from your home to the town centre?_____

2 You need to be at the town centre by 2.05 pm (1405). You have to travel by bus. Which is the **latest** bus you can catch from your home to be at the town centre in time? _____

Here is the timetable for bus 33.

Timetable

Stops	Journeys											
	A	B	C	D	E	F	G	H	I	J	K	L
2 Town centre	0805	0905	1005	1105	1205	1305	1405	1505	1605	1705	1805	1905
5 West Stadium	0830	0930	1030	1130	1230	1330	1430	1530	1630	1730	1830	1930
5 West Stadium	0835	0935	1035	1135	1235	1335	1435	1535	1635	1735	1835	1935
2 Town centre	0900	1000	1100	1200	1300	1400	1500	1600	1700	1800	1900	2000

3 How many **journeys** are shown on route 33 timetable?_____

4 How many **stops** are there on each journey?_____

5 Which **places** are linked by bus 33? _____ and _____

6 **How long** does the journey between the two places take? _____

7 Mark the route of **bus 33** on the map in **blue**.

8 If you need to get to West Stadium before 3 pm (1500), which is the **latest** bus you can catch from the town centre?_____

Name:

In time for United
(continued)

Here is a timetable for the third bus, number 121. It is a circular route which goes one way round the town.

Stops	Journeys									
Timetable										
	A	B	C	D	E	F	G	H	I	J
1 Home	0800	0900	1100	1200	1300	1400	1500	1700	1800	1900
4 East Side	0845	0945	1145	1245	1345	1445	1545	1745	1845	1945
3 South Bank	0930	1030	1230	1330	1430	1530	1630	1830	1930	2030
5 West Stadium	1015	1115	1315	1415	1515	1615	1715	1915	2015	2115
1 Home	1055	1155	1355	1455	1655	1755	1855	1955	2055	2155

What to do on the map

On the map on page 40:
● Shade the route of **bus 121** in **green**.
● The four stops are shown and named. In the boxes near each stop, write in the **times** when the buses **leave** that stop.
● In each circle by the boxes, fill in the time taken by the bus to reach that stop from the previous one.

Getting to the match in time

You want to see the match at West Stadium. You have to get there, and home again, by bus. The match begins at 3 pm (1500) and finishes at 4.40 pm (1640).

What to do in the table

● Complete the table below. This will be your personal timetable telling you which buses to catch, when to catch them, and where to catch them.
● Do it in the order shown by the numbers in the left-hand column.

	Journey	Bus route	Time
4	Leave home		
3	Arrive at town centre		
2	Leave town centre		
1	Arrive at West Stadium		
	Match begins	✕	1500
	Match finishes	✕	1640
5	Leave West Stadium		
6	Arrive home		

● Now shade the boxes marked 'Bus route' and 'Time' in the same colours as those used on the map to shade the routes. These are bus **62** in **red**, bus **33** in **blue**, bus **121** in **green**.

Ride round Weston

There are details of a bus route below the map of Weston on this page.

What to do

● Look at the map and then read the description of the bus route.
● Using a **red** pencil, mark the **bus route** on the map. Show this colour on the **key**.

Key

B	bank
BD	bus depot
C	coffee house
G	garage
Jc	Jobcentre
N	newsagent
O	Balfours' home
PO	post office
S	junior school
SM	supermarket
T	telephone
—	bus route
▲	bus stop
✝	St Mary's Church

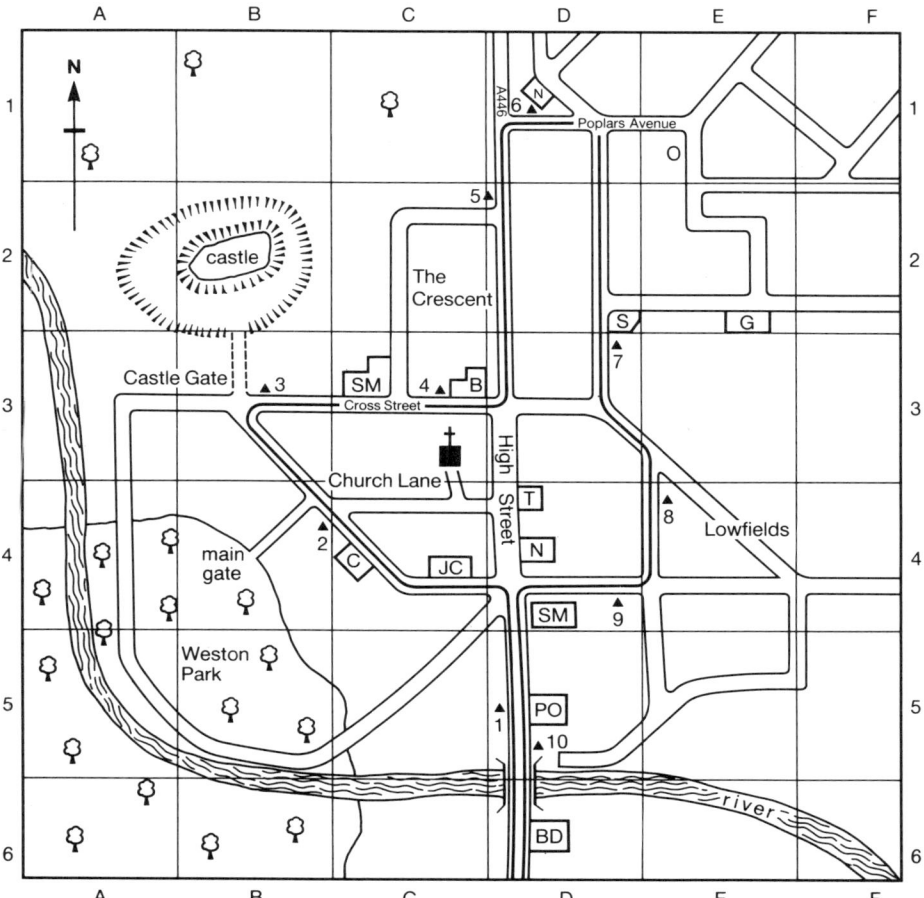

Bus route

The bus leaves the **depot** south of the river and turns north across the bridge. The first stop is opposite the **post office (grid square D5)**. The bus then continues north and turns west at the Jobcentre. The second stop is at the western end of **Church Lane (B4)**. The bus proceeds to the third stop **Castle Gate (B3)** in Cross Street. The fourth stop is at the **bank** on **Cross Street (C3)**. The bus then turns north along the **A446**. The fifth stop is at the junction with the **Crescent (D2)**. The bus goes north again and then turns east into **Poplars Avenue (D1)**, where the sixth stop is reached. From there the bus goes south to the seventh stop, near the **junior school (D2)**. The eighth stop is in square **E4** and after this the bus continues south again to the **crossroads in E4**. Then it turns right to the ninth stop **(D4)** near the **supermarket**. It turns south on reaching the High Street, and arrives at the tenth stop at the **bridge (D5)**.

Ride round Weston
(continued)

Make a **pie chart** to show how many buses end their journeys at particular bus stops.

Databank: bus timetable

Bus stops	Bus stop number	Bus journeys										
		A	B	C	D	E	F	G	H	I	J	K
Post Office	1	0800	0830	1000	1200	1400	1600	1630	1700	1730	1900	2200
Church Lane	2	0805	0835	1005	1205	1405	1605	1635	1705	1735	1905	2205
Castle Gate	3	0810	0840	1010	1210	1410	1610	1640	1710	1740	1910	2210
Bank	4	0815	0845	1015	1215	1415	1615	1645	1715	1745	1915	2215
Crescent	5	0820	0850	1020	—	1420	1620	1650	1720	1750	1920	2220
Poplars Avenue	6	0825	0855	1025	—	1425	1625	1655	1725	1755	1925	2225
Junior School	7	0830	0900	1030	—	1430	1630	—	1730	1800	1930	2230
Lowfields Estate	8	0835	0905	1035	—	1435	1635	—	1735	1805	1935	2235
Supermarket	9	0840	—	1040	—	1440	1640	—	1740	—	1940	2240
Bridge	10	0845	—	1045	—	1445	1645	—	1745	—	1945	2245

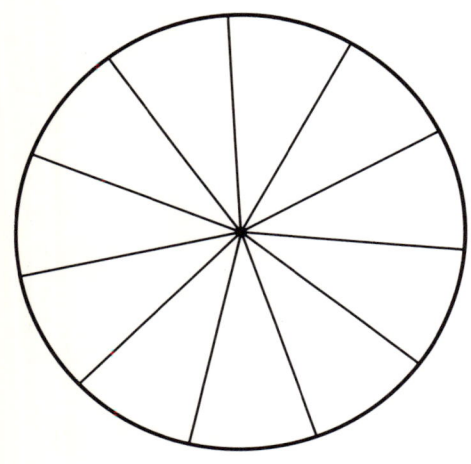

What to do

Look at the timetable above.

1 How many buses, starting at the post office, end their journey at the **bridge (stop 10)**?
▶ Shade that number of sectors on the pie chart **red**.

2 How many buses, starting at the post office, end their journey at **Lowfields (stop 8)**?
▶ Shade that number of sectors on the pie chart **orange**.

3 How many buses, starting at the post office, end their journey at **Poplars Avenue (stop 6)**?
▶ Shade that number of sectors on the pie chart **yellow**.

4 How many buses, starting at the post office, end their journey at the **bank (stop 4)**?
▶ This number should be the same as the remaining number of sectors on the pie chart. Leave these **unshaded**.

Getting to work

Look at the timetable above and the map of Weston on page 43.

Imagine you live opposite the post office and work at the bank.

You start work at 0915 (9.15 am) and catch a bus from the post office to get there.

5 At what **time** would you **catch** your bus?

6 At what **time** would you **get to work**?

7 **How long** would you have to wait **before starting work**?

You finish work at 1700 (5 pm).

8 **Which bus** would you **catch** to your friend's house in Poplars Avenue?

9 At what **time** would you **reach** your friend's house?

Satelvision

This is a game for two players.

Each player is the owner of a major TV company which has launched a number of satellites into space. With these satellites each company can reach all the areas shown on the map on page 48.

The map is divided into shaded and unshaded squares. From the key, you can see which areas have many TV sets and which have few.

What you have to do is to try to beam into the largest TV audience. That is, try to get as many shaded squares as you can, so that you get a big score. The values of the squares are given in the key and also in the scorecard on page 48.

The winner is the player with the highest score when the scorecard is completed.

How to play

1 You will need a die. Each of you will need a different coloured pencil and **your own record sheet**. You both use the **same map**.
2 Spin a coin to decide who will have the first throw of the die. During the game, the first player will answer questions 1 – 30 on page 47. The second player will answer questions 31 – 60.
3 The first player throws the die, crosses out number 1 on the throw line of the databank and writes the number thrown in the box.
4 The other player then throws the die. If he/she throws **2, 4** or **6**, the **first player** must follow instruction **A** in the box on the right. If he/she throws **1, 3** or **5**, the **first player** must follow instruction **B**.
5 The second player now throws the die again and follows **instruction 3**.
6 The first player then throws the die and **both** players follow **instruction 4**.
7 The game goes on in this way until either **all the squares** have been shaded **or** both players have written a number in **every box** on the **number thrown** line.
8 If your opponent runs out of challenge questions, then you simply shade in the number of squares as shown by the number on the die when you throw it.
9 At the end of the game, count the squares won by each player and complete the scorecard.

A

Answer a challenge question from your list on page 47.

The page in the book where the answer can be found is given in brackets after the question.

If you answer correctly, you can shade in some squares on the map. The number of squares you can shade is the same as the number you threw on the die and wrote in the number thrown box.

Do not shade in any squares if your answer is wrong. Your opponent can check that you have given the right answer. If you are both unsure, ask your teacher.

B

You do not have to attempt a challenge question. You can shade in some squares on the map.

The number of squares you shade is the same as the number you wrote in the number thrown box.

Record sheet

Player's Name _____

Database																		
Throw	1	2	3	4	5	6	7	8	9	10	11	12	13	14	15	16	17	18
Number thrown																		
Throw	19	20	21	22	23	24	25	26	27	28	29	30	31	32	33	34	35	36
Number thrown																		
Throw	37	38	39	40	41	42	43	44	45	46	47	48	49	50	51	52	53	54
Number thrown																		
Throw	55	56	57	58	59	60	61	62	63	64	65	66	67	68	69	70	71	72
Number thrown																		
Throw	73	74	75	76	77	78	79	80	81	82	83	84	85	86	87	88	89	90
Number thrown																		

Answer your questions on the lines below.

If you answer correctly, shade the box at the end of the line.

1 _____ ☐   16 _____ ☐

2 _____ ☐   17 _____ ☐

3 _____ ☐   18 _____ ☐

4 _____ ☐   19 _____ ☐

5 _____ ☐   20 _____ ☐

6 _____ ☐   21 _____ ☐

7 _____ ☐   22 _____ ☐

8 _____ ☐   23 _____ ☐

9 _____ ☐   24 _____ ☐

10 _____ ☐   25 _____ ☐

11 _____ ☐   26 _____ ☐

12 _____ ☐   27 _____ ☐

13 _____ ☐   28 _____ ☐

14 _____ ☐   29 _____ ☐

15 _____ ☐   30 _____ ☐

Name:

Challenge question sheet for Satelvision

Questions 1 – 30 for first player

1 What can best be done with fertile land? (2)
2 How many people live in the world? (3)
3 Name one country west of the Greenwich Meridian which is least crowded? (3)
4 Is Britain north or south of Nigeria? (4)
5 Is Mexico south of the USA? (4)
6 Name one country west of the Greenwich Meridian which is not very crowded? (3)
7 Are the most crowded countries mainly north of the Equator? (3)
8 Which country in Great Britain has the greatest number of most crowded counties? (5)
9 Is London one of the very crowded counties? (5)
10 What word best describes areas which are too hot or too cold? (6)
11 Which country in Great Britain and Northern Ireland has the smallest population? (10)
12 What is the population of Africa? (11)
13 Does the USSR have a population of 70 million? (11)
14 Does the continent with the largest population also have the largest amount of cropland? (15)
15 How much cropland is there in Europe? (14)
16 What is the average temperature in London in January? (20)
17 In which month does London have its highest average temperature? (20)
18 What is the longitude of London? (21)
19 What is the average temperature in Athens in July? (22)
20 Which two months in Athens have the lowest average temperature? (22)
21 How many degrees warmer is Athens than London in July? (25)
22 Is Athens always warmer than London? (25)
23 Is Athens 10 °C warmer than London in July? (25)
24 In which place on page 29 could you visit a museum? (29)
25 Is London, on average, warmer than Stavanger? (32)
26 Is the average August monthly temperature in Lisbon 23 °C? (32)
27 Which type of holiday was least popular? (34)
28 Name one town in Spain? (37)
29 Is the home north of the town centre on page 40? (40)
30 Is the supermarket in square C4 on page 43? (43)

Questions 31 – 60 for second player

31 What type of land is in square 0273 on page 2? (2)
32 Name one country east of the Greenwich Meridian which is least crowded? (3)
33 Is China east or west of Britain? (4)
34 Are the most crowded countries mainly east of the Greenwich Meridian? (3)
35 Name a country in Great Britain which has none of the most crowded counties? (5)
36 Which North American country is in two parts? (6)
37 Which country is north of England? (10)
38 Does Africa have a greater population than Europe? (11)
39 Which country has the most people living in it? (12)
40 Do more people live in the USA than in the USSR? (12)
41 Which two continents have the same amount of cropland? (14)
42 Which continent has a very small population total and a cropland of over 9 million square kilometres? (15)
43 What is the warmest monthly temperature in London? (20)
44 In which month in spring does London have its highest average temperature? (20)
45 What is the average monthly temperature in Athens in December? (22)
46 Is London always cooler than Athens? (23)
47 What country is Athens in? (22)
48 How much warmer is Athens than London in January? (24)
49 What would you do on a holiday in south-west England? (29)
50 Is the weather important for all of Blackpool's holiday activities? (29)
51 Is Lagos always hotter than Lisbon? (32)
52 Is Lagos south of Lisbon? (32)
53 Is Stavanger south of London? (32)
54 What is the average temperature of Lagos in March? (32)
55 What type of graph is usually used to show rainfall? (38)
56 Were holiday camps more popular than self-catering holidays? (34)
57 Was good weather important to many people when they were on holiday? (35)
58 Is Madrid next to the sea? (37)
59 Is there much rain in Malaga in July? (38)
60 Is the church west of the High Street on page 43? (43)

Name:

Satelvision map

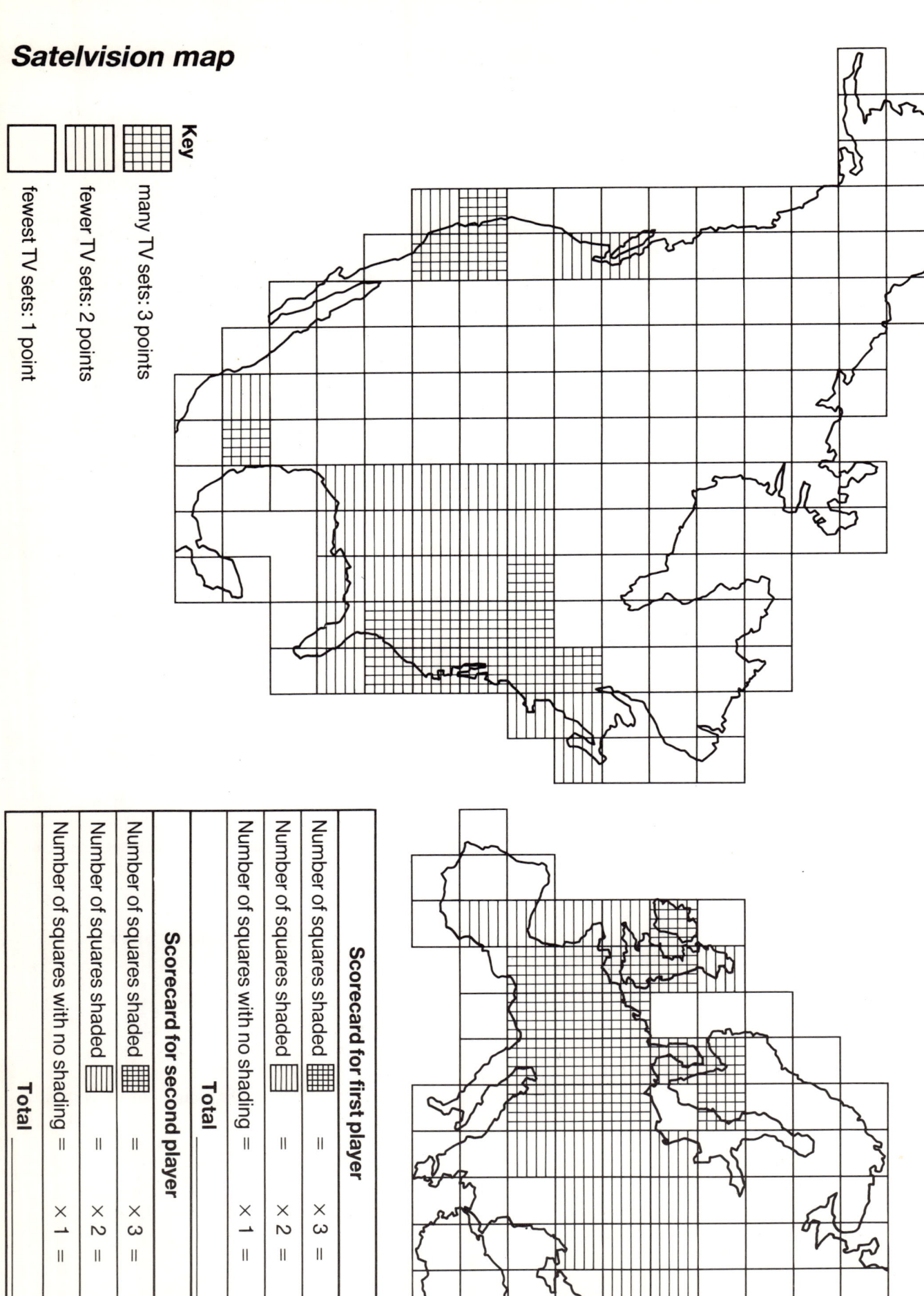

Key

many TV sets: 3 points

fewer TV sets: 2 points

fewest TV sets: 1 point

Scorecard for first player

Number of squares shaded ⊞ = × 3 =

Number of squares shaded ▥ = × 2 =

Number of squares with no shading = × 1 =

Total

Scorecard for second player

Number of squares shaded ⊞ = × 3 =

Number of squares shaded ▥ = × 2 =

Number of squares with no shading = × 1 =

Total